THE LAYMAN'S GUIDE TO MAMOD STEAM ENGINES

Dedicated to every model live steam enthusiast out there that is keeping this hobby alive.

© RestoringMamods.com - William Green 2014

All rights reserved. No part of this publication may be reproduced, stored in a retrieval system or transmitted, in any form or by any means, electronic, mechanical, photocopying, recording or otherwise, without the permission in writing from the publisher.

First publicly published in September 2014

Photograph Credits

All images copyright of RestoringMamods.com with the exception of the following:

Mikes-Steam-Engines.co.uk: Hobbies/Mamod SE4 P9, SE2/ Grinding Wheel/ Minor 2. Camden Street P10, SE2 Mazak/ SE2 Raised Base/ SE3/ ME1/ SR1 P11, TE1/ TE1a/ SR1a P12, SA1 P13, SA1L P15, MM1 (left, middle and right), MM2 (P21), SE1 (left, middle and right)/ SE2 (left and middle)/ SE3 (middle) P22, SE4 (left and middle)/ SP1 (left)/ SP2 (left) P23, ME3/ ME1 /TE1/a (left, middle and right)/ SR1/a (left, middle and right) P26, SW1 (left and right)/ SA1 (left, middle and right)/ SA1L (left) P26, FE1 (left) P28, London Bus (left and right) P28, TWK1 (left)/ RS1 P29.
Ian's Online Postings from indianarog.com: Minor 2 P9.
Erik from Steam-Toys.com: Meteor P10.
LiveAuctioneers.com: Conqueror P10, SW1 P13.
Wikipedia.org: ME2 P11.
Mike Jane, Plymouth: MEC1 P12, G&G SE3 P13, SE3 (left) P22, SP8 (left and middle) P25, ME2 P26.
Mick Wilde: SE2a P12.
SteamCollector.com: Open Wagon P13.
ArrowModels.com: Fuel tablets P13.
Academic.ru: SP1 P14/23.
Tls1963: SP3 P14.
Forest-Classics.co.uk: SP4 P14/24 (left), SP6 (middle and right) P25, P102 (bottom).
ManorModels.co.uk: SP5 P14/24 (left), P72, P73, P88, P89 (right), P102 (top).
Mamod.co.uk: WS1 P14/ TWK1 P14, Delivery Van/ London Bus P15, SP5 P16/24, SP5D/ Gas Burner P17, SP6/ Centurion/ Challenger/ Showman's Special P18, Mark 2/ Mark 1/ Brunel Vertical Engine/ Diamond Jubilee Saddle Tank/ SP8 P19, SP6 (left) P25, SA1L (right) P27, TWK1 (right)/ Vertical Brunel (left) P29, P94.
The Station Masters Limited: RS1 P14.
Courtrade.com: RS2 P14.
Dougrail, West Midlands: SLK1 P15.
Vectis Auctions Ltd: Brass Roadster P15, MM2 (right) P21, SA1L (middle) P27.
Mamodonline: FE1 P15.
SteamReplicas.co.uk: Le Man Racer/ William Locomotive P17, Delivery Van (middle) P28.
BullyBeef.co.uk: Le Man RC Racer/ 'O' Gauge Track P17.
Quirao.com: Harry The Rocket P18.
Rory Williams: SE4 (right) P23, SE3 P24.
David Walmsmey: SE1/2 P9.
Piston42 from BloOOo.fr: SE3 P9.
Ralph Laughton: SP3 (right) P24.
Online Model Steam Engine and Aeroplane Engine Store: SP5 (middle) P24. FE1 (middle and right) P28.
Wikimedia.org: SP7 (left) P25.
Photobucket: SP7 (middle) P25.
PLAYING AROUND! Via YouTube: SP7 (right) P25.
Ministeam: SP8 (middle) P25, London Bus (middle) P28.
AnticsOnline.co.uk: Delivery Van (left) P28.
Tennants Auctioneers: Delivery Van (right) P28
emhretail at eBay Inc.: TWK1 (middle) P29.
Courtrades.com: RS2 P29.
Gyroscope.com: Vertical Brunel (middle) P29.
Chris Cairns: Vertical Brunel (right) P29
Wonderland (Toys) Limited: D305 (left) P30.
QCTechnologies via YouTube: D305 (right) P30.
Online Safety Program For Truck Drivers: P33.
Furred Pipe: P49.
John Daniel: P55.
MATTE, The University of Liverpool: P71.
Andrew Gorton: P74 (all).
Model Enthusiasts: P103 (top).
Moo Ltd.: P103.
eBay Inc: P104.

Designed and produced by Will Green.
Proofread and edited by Phil Green and Jane Green.

Website: www.RestoringMamods.com
Email: bloggeradmin@restoringmamods.com

IBSN: 978-1-326-09676-2

THE LAYMAN'S GUIDE TO MAMOD STEAM ENGINES

Purchasing, Restoring, Repairing and Maintaining your Mamod Steam Engine … plus a few extras inside

William Green

"Any problem can be fixed – it is just a matter of time, money and dedication"

Contents

Introduction and Acknowledgements	6
The History of Mamod	8
- Mamod Timeline (1936-2014)	9
The Range of Mamod Models	21
Steaming Your Engine Up	32
- Preparations and Safety	33
Buying Your Engine	40
Clues to a Damaged Engine	48
How the Steam Engine Works	51
General Look - Mamod Stationary Engines	57
General Look - Mamod Mobile Engines	61
Equipment/Tools Needed For a Restoration	64
Restoration No Nos!	65
Universal Restoration Tips	67
Restoring a Mamod Stationary:	
- Restoring a Mamod SE	76
- Restoring a Mamod SP	80
- Restoring a Mamod Minor	84
Restoring a Mamod Mobile:	
- Restoring a Mamod TE Traction	87
- Restoring a Mamod SR Steam Roller	92
- Restoring a Canopy	94
- Restoring a Lumber Wagon Trailer	96
Problems with Starting Your Engine	98
Buying Parts for Your Engine	102
Glossary	105

Introduction

This book contains everything that you need to know when it comes to purchasing, restoring and maintaining your beloved Mamod steam engine along with a few extras. As you might already know, I am the owner and creator of RestoringMamods.com where I publish details of all of my Mamod restorations. Although the articles on RestoringMamods.com are useful to model steam engine enthusiasts, they do not tell people the details of just *exactly* how to restore these types of engines. Through the experience I have gained over the years from restoring such engines, I have created this book to share with you everything I know.

I strongly believe that some of the best engines are the earlier models from the past. The Mamod range, in general, is efficiently simple and enables anyone to learn the basics of how an external combustion engine worked before the internal combustion engine. They are cheap to buy, refurbish and provide hours of enjoyment. There is nothing like the smell of a Mamod engine in full working glory!

To make things easy, I have split the book into key sections looking at all the stationary engines and some of the traction engines out there. These range from:

- The Mamod SE range.
- The Mamod SP range.
- The Mamod Minor range.
- The TE and SR range
- The LW1 Lumber Wagon

If the engine you are wanting to restore is not listed, fear not! The restoration process for all Mamod engines is very similar. Therefore, you can still gain a lot of valuable tips from looking at the restoration process of similarly built engines to the one you own.

Together, with a detailed explanation of how to restore Mamod engines, I have also included some general historical background to the brand 'Mamod' with the full range of engines they have engineered since the brand was first created. I highly recommend you read this section since it might interest you in purchasing more engines. You will be surprised by the range of engines Mamod have produced over the years!

Using This Book

There are many ways you can use this book to your advantage. If you want to gain a full understanding of Mamod engines, you can read through the whole book. However, this book works best by your side, much like a manual, whilst working on your engine. It is dual purpose and works as a great companion at being a book for when you need it.

At the back of the book is a glossary of all the key terms I have used throughout. Therefore, if you are unsure about any terms used in this book, please refer to the glossary.

Acknowledgements

As the author of this book, I would like to give a special thanks to my parents, Phil and Jane Green. My father, who has been a companion in helping restore some of my engines and proof read and edited this book. Without his help and dedication, some of the engines would have been much more difficult to restore. My mother for her patience, proof-reading and editing skills!

As well as this, I would also like to thank:
- Mamod for providing key information to help me form a historical timeline of Mamod and their models.
- Andrew Gorton from 'down under' (to me) for allowing me to use some of his photos and provide a great detail of help and advice.
- Michael for allowing me to use his pictures from his website mike-steam-engines.co.uk.
- James Grainger for proof reading different areas of the book.
- Lastly, Dave Cramphorn, a true steam engine enthusiast!

William Green July 2014

My SW1 Steam Wagon and TE1a Traction Engine on a nice spot of grass on a steam up day.

History of Mamod

Mamod is a British toy manufacturer that produced model steam engines. The company was founded by Geoffrey Malins in 1937 in Birmingham. During the early years of the Company, the steam engines were sold under the 'Hobbies' brand name. However, after a short while, the brand name changed to Mamod, deriving from the 'Malins Models' name - being shortened to Mamod ('Ma'lin + 'Mod'els = Mamod). Many of the earliest models were hand-built by Geoffrey Malins himself, making them extremely valuable.

Mamod steam engines were aimed at the toy market since they were simple in design and ran at low boiler pressures to make the engines deemed safe for children to use. All of the engines work through an oscillating cylinder. Most of the engines' speed was uncontrollable. However, on the more complex engines, regulators or forward/reverse levers were fitted to limit the flow of steam to the piston, hence altering the speed and direction of the piston oscillation.

> *Did you know?*
> *Mamod, the British toy manufacturer, is over 75 years old and survived through World War II!*

The first engines produced by Mamod were stationary engines. This consisted of the SE (stationary engine) range which featured a boiler mounted onto a firebox. The boiler was then connected to the piston/s through copper piping. Although these models of engines were produced over 80 years ago, the whole structural design of the SE range has not changed significantly over the years.

To further improve the safety of the engines, Mamod changed the type of fuel used in the mid-1970s. It was originally either a multi-wick burner or a vaporising meth burner which changed to solid fuel tablets. This was a huge improvement in safety for the Mamod range since methylated spirit was used for vaporising burners. Methylated spirit has an invisible flame and can easily be spilt from the burner onto the child. The solid fuel tablets of Mamods have a colour flame, thus improving their appeal to the younger audience.

It was apparent that the takeover by Charlie Cooper was one of the toughest times of Mamod's history in the early 1980s. Sales were decreasing and Mamod was finally rescued by David Evans who introduced the TWK1 Tractor and Wagon kit. This was a breakthrough for a failing company since the TWK1 Tractor used parts from other Mamod models to minimise production costs. The front wheels were from the Roadster while the flywheel was from the Steam Wagon and the canopy was removed completely. This can be seen to represent the rebirth of Mamod.

What follows is a timeline of Mamod (specifically Malins' development with Mamod) from 1936, the very start, to the present day. However, if you want a more in depth look at Mamod's history, I strongly recommend purchasing 'The Story Of MalinsModels' book - even if it is very expensive!

Mamod Timeline (1936 -2014)

1936 Geoffrey Malins teams up with Hobbies to supply engines. He produces 576 SE1/2/3/4 engines for Hobbies in his first year

Hobbies SE1

Hobbies SE2

Hobbies SE3

Hobbies SE4

Mamod Minor 2

1937 Mamod is born as Geoffrey continues to build engines for Hobbies. He also starts to build Mamod engines too. The SE4 is officially born. Geoffrey describes the engines he first built as *'the worst I ever made as I had to begin at the beginning and find out everything'*.

1938 Malins moves to a larger workshop, 2 St Mary's Row in Birmingham, to build even more engines. He is joined by his son, Bud, and 30 more part-time workers. Mamod is now a fully running business.

1939 Mamod officially becomes a private limited company named Malins (Engineers) Limited and has a four page brochure to display the different engines Mamod has to sell. As well as this, the first Minor 2 engine was introduced. It had twin horizontally opposed cylinders and was discontinued very shortly after being introduced in 1940. It is one the rarest engines since only a few dozen of these engines were made.

1940 - 1945 World War II (started 3rd September 1939) decreased the total production of steam models considerably. The Board of Trade gave Geoffrey permission to produce up to £200 (£9,750 in today's money) worth of engines per month, whilst carrying out other work for the war effort.

1946 The war is over (ended 8th May 1945) and Mamod slowly starts to recover, only producing Mamod SE1s and SE2s. Production resumes back to normal.

Malins creates two prototype SE4 engines which were never put into production. This SE4 prototype engines Malins created are looked upon as the 'Holy Grail engine' as only two were ever produced. Malin's son, Eric, becomes the head of production after being demobbed from the Royal Navy.

Mamod SE4 Hand built by Geoffrey Malin.

There is only two of these engines in existance.

Next page 1947-52

1947 Another of Malin's sons, Phil, joins the family business after being demobbed from the Royal Indian Army, with the rank of Major. Short supplies of materials and unreliable equipment makes production post-war slow. Mamod becomes the most popular British steam toy maker since the public was put off German steam toys.

1948 Production picks up with the introduction of the brass flywheel to the SE2, the Grinder accessory and a new improved Mamod MM2 engine. Eric Malinsis now in charge of production and costing.

1949 Mamod's production moves to larger premises at 25-31 Camden Street. This is also the year Mamod introduces their first ever steam engine boat, the 'Meteor' which can be considered to be the Company's only ever commercial failure.

Mamod Meteor ⬇

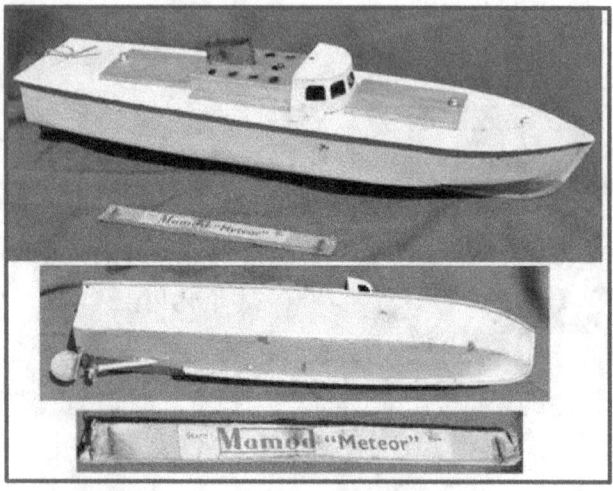

Mamod at Camden Street ⬆

1950 The Meteor is not selling well at all especially at the princely sum of £4 5s 0d (around £120 in today's money). To save the Meteor, another boat engine was introduced named the 'Conqueror' which was powered by a FROG 'Revmaster' electric engine.

1951 Mamod has fully moved and settled at Camden Street. The Meteor is still selling but in small quantities. Therefore, to help the company through the quieter months, Mamod sold solid brass ball-catches which, to this day, is still the only non-toy item Mamod has ever sold.

1952 A dismal year for Mamod as Meteor's production stops at 1,500 units and the Conqueror at 200 units. These boats actually lost the company £10 with every engine sold. Bud Malinsleaves the company due to disagreements with his father.

Next page 1953-61

1953 Innovations occur on Mamod's engines. The brass flywheel changes material to Mazak which also starts to be used on the tools too. As well as this, it is the last year that the flat base stationary engines were created.

Mamod SE2 featuring a Mazak flywheel.

1954 The stationary engines have a base now made of pressed steel. Eric Malins walks out of the company. Due to this, Geoffrey Malins increases his workforce to 40 part-time employees to ensure the company continues to run smoothly.

Mamod SE2 with a raised base.

1955 Geoffrey Malins looks for someone to buy his now-profitable company. He strikes a deal with Eric who gets the overall control of Malins Engineers Ltd with Geoffrey becoming Chairman.

1956 Eric Malins is the Managing Director of Mamod and looks to create a new engine which will be Mamod's first new model engine in nigh on seven years.

1957 Mamod introduces the SE3 engine with twin cylinders which is considered to be one of Mamod's best engines. Since the SE3 uses a vaporising burner, the SE3 sparks the end of the multi/single wick burners.

Mamod SE3 with the updated vaporising burner.

1958 Two more engines are introduced being the Mamod ME1 and ME2. As well as this, the firebox design is changed on the SE1/2 and MM2. The only engine to use a wick burner is the MM1 with all others using vaporising burners.

Mamod ME1

Mamod ME2

1959 The 1950s have been good for Mamod: their only real competitor was SEL. The 1960s was looking a promising decade for them.

1960 The workforce is now 50 employees larger producing 300 engines (excluding accessories) a day. Camden Street was becoming a bit too tight for Mamod's production. Therefore, expansion of premises would be needed in the 1960s.

1961 This would be Mamod's last year at Camden Street. More importantly, Mamod launch the SR1 Steam Roller which was Mamod's first real mobile engine. It was a huge success and was the start of a great range of mobile engines.

Mamod SR1

Next page 1962-69

1962 As demand for the SR1 steam roller increased, production is moved to 206 Thorns works in the West Midlands (the old pipes works at Brierley Hills). It was purchased at a cost of £10,000 and found by Geoffrey Malins (in today's currency around £180,000). This was a great success for Mamod who were beginning work on their second mobile engine.

1963 This was a great year for Mamod as they launch the most successful engine in the Company's history. This was the biggest-selling model ever produced by the company: the Mamod TE1 Steam Tractor.

Mamod TE1

1964 The TE1 becomes Mamod's bestselling engine along with the SR1 Steam Roller.

1965 With production costs rising, Mamod phased out the use of screws on their engines and started using pop rivets. In celebration of this change, Mamod launched a new engine in collaboration with Meccano: the MEC1.

Mamod MEC1

1966 Mamod continues to progress with some more future innovations developing: this time, in the mobile engines. As well as this, in an attempt to cut costs, Mamod began to produce pressure die-castings in-house. A machine stop and paint shop were also added to the premises.

1967 The 'a' series is introduced bringing forward the TE1a and SR1a which both have forward/reverse levers to control the direction of the engine and its speed. The stationary engines are improved too bringing in the SE1a and SE2a while the Minor range remains unchanged.

Mamod TE1a   Mamod SR1a

Mamod SE1a   Mamod SE2a

1968 The SR1 Steam Roller loses its aluminium rolls to Mazak. New floor space at Thorns Works is ready to accommodate new machinery.

1969 The new floor space at Thorns Works makes it possible to die cast parts, slashing the cost of production. Mamod introduces the OW1 open wagon and LW1 lumber wagon for the TE1a and SR1a to pull. Steam engines are introduced to schools with the Mamod Griffin and George SE3 engine - this engine is hard soldered. As well as this, Steve Malins joins the family business straight from school. Mamod wins their first NATR award.
(Pictures of SE3 and wagons on the next page).

Next page 1970-77

Mamod Griffen and George SE3

Mamod Lumber Wagon

Mamod Open Wagon

1970 The Mamod Minor 1 gets a vaporising burner while Eric Malins accepts the second Company's award (first in 1969) from the National Association of Toy Retailers.

1971 Steve Malins wants his 'first model' to be manufactured by Mamod: the Steam Road Wagon.

1972 The SW1 Steam Wagon (by Steve Malins) is launched. All engines with a whistle now have the finger sprung whistle fitted as Mamod wins their second award from the NATR.

Mamod SW1

1973 Production hits nearly 116,000 units.

1974 David Evans joins the company with the objective of introducing a production and quality control system.

1975 The creator of Mamod, Geoffrey Malins, dies aged 83. Mamod's production is hit hard as orders are cancelled due to concern over the fuel used because of an 'accident' occurring in America. The incident involved methylated spirit highlighting the dangers of the fuel that Mamod models use. Some valued staff members are made redundant as a result of journalism lunacy. Mamod wins another NATR award.

1976 The SA1 Steam Roadster, Steve Malins' engine, is launched and becomes an instant success. This helps the company turnaround from 1975 which was considered not one of Mamod's best years.

Mamod SA1

1977 After the incident of 1975, liquid fuels are banned. All of Mamod's engines now come with solid fuel and solid fuel burners. The fuel is made at the Thorns Works. As well as this, Malins (Mamod) Limited was formed.

Mamod Solid Fuel Tablets

Next page 1978-82

1978 This is the last year the SE range was built with the following year having a totally new designed stationary range. As well as this, to accommodate new European regulations, all engines, except the Minor 1, now have sight glasses instead of water level plugs on the boiler. Surprisingly, the SA1 Roadster sold very well in Germany as there was no other model like it in the German market. This enabled Mamod to enter the German market with more of their engines.

1979 Mamod introduced locomotives at a steam toy fair. They received high praise for them from trade-buyers who saw them for their enormous potential. However, Mamod did not see the attraction of these engines. Yet, they were destined to become one of the most exciting toys they ever created.

The new SP range is introduced featuring five stationary engines: SP1/2/3/4 and the twin cylinder SP5. As well as this, Mamod also introduces the WS1: all of the workshop tools on one base. The SP range was introduced to refresh the range of models the family-owned Company made, because Mamod thought the SE range was *'looking tired'* and had to keep up with new safety regulations.

Mamod SP1

Mamod SP2

Mamod SP3

Mamod SP4

Mamod SP5

Mamod WS1

1980 The Steam Train range is introduced with the RS1 and RS2. The banking profession 'brings down' Britain's most successful family run toy steam company. The Malins family no longer owns Mamod: Mamod now has a new owner being the bank which then sold the Company onto a local businessman named Charlie Cooper. He liquidated both Malins(Engineers) Limited and Malins(Mamod) Limited to create a new company: Mamod 80 Limited.

Mamod RS1

Mamod RS2

The Malins family built Mamod from nothing to the biggest toy steam company in Britain. Every owner of a Mamod engine should appreciate the Malins Company for how they have provided generations with a long and enjoyable legacy of toy steam engines.

1981 Efforts to boost sales were diminished when Charlie Cooper opted to increase all prices by 27.5%. In the end, staff were informed that all remaining stock had been sold to BTP (Ascot) Limited and the Thorns Works production was to end resulting in just two staff members joining the new factory in Ascot. Mamod Limited was formed.

1982 Production resumed and the first model kit was introduced being the TWK1 tractor and wagon kits despite the lack of finance of Mamod's new owners BTP. In all the years of Mamod being a company, the Malins family had never liked the idea of a model kit so this was a big move by the new owners.

Mamod TWK1 Kit

Next page 1983-89

1983 After success from the TWK1 kit, Mamod introduced another kit: the SLK1 kit which was the first railway locomotive that was made available in kit form. Mamod also introduced a special edition engine being the Brass-finished Roadster (production of these was limited to just 1,170).

Mamod SLK1

Mamod Brass Roadster

1984 The first variation of the Roadster was introduced being the SA1L (was the first of many variations) and was created with the Rolls-Royce Silver Ghost in mind. Kits were made for the SA1 and SA1L. Everything seemed to have been going well for BTP and Mamod. However, BTP lost one of its largest contracts for motor components and the collapse of the De Lorean meant they were badly in debt. This left BTP no choice but to sell the Mamod division to Jedmond (Engineers) Limited who were part of the Starwest Investments in Berkshire. Production ceased during the takeover.

Mamod SA1L

1985 Production resumed in June at the Jedmond (Engineers) Limited new works. Mamod discontinued some products that had poor sales being the SP1, SP3, SP5 and many of the accessories built for the stationary engines.

1986 The FE1 was introduced. Like the SA1L, it was a variation of the Roaster which featured its chassis and running gear to minimise the cost of this new engine.

Mamod FE1

1989 It is Mamod's 50th Anniversary (if you count the start of Mamod to be 1939).

The Delivery Van and London Bus were introduced to the line-up of engines on sale. These two engines, like the SA1L and FE1, were again variations of the roadster.

Mamod Delivery Van

Mamod London Bus

Next page 1990-99

1990 The company changes hands again from Starwest Investments to Adam Leisure Group PLC. This caused production to pause as it was moved to a small factory unit in Blaydon near Newcastle-Upon-Tyne.

1991 Adam Leisure Group PLC was taken over by Porter Chaburn PLC just as production started up again. Be that as it may, there were unsuccessful talks to move production as far afield as Hong Kong. This led the company to selling Mamod in 1992.

During this difficult time, Mamod had been very intelligent in the sense that they managed to continually introduce new products without having to completely build new products. Since the company had the objective of keeping costs to a minimum because they were struggling financially (especially with the number of moves they have had in ownership), it was a great idea to make variations of engines to suit different consumer needs. This was not possible with the Tractor but, as you can see, the variations from the Roadster were a huge success for the company which helped considerably to keeping Mamod alive.

1992 After approaching Thomas Johnson & Company, who supplied pressings to Mamod for many years, the company once again had ownership. Production was moved to Erdington in Birmingham which coincidentally was a few hundred yards from where Geoffrey Malins started his very first business. All of Mamod's assets to Birmingham was completed by 1992. In May of 1992, production was ceased but resumed again in October.

1996 In the past ten years, Mamod had no fewer than seven different owners and five different homes. Mamod has new owners now, Peter and David Terry. They agreed the Erdington works were too large for their needs so decided to move production to Smethwick in the West Midlands, a smaller factory, in May.

1992-1997 During this time, all of Mamod's efforts were concentrated on refurbishing the worn tooling at the factory and on improving the quality of the engines being made.

1999 The SP5 twin cylinder stationary engine was reintroduced. New base plates were made with a larger flywheel added and instead of a brass chimney a die cast chimney was used. The engine could only move work in one direction.

Mamod SP5

Next page 2000-04

2000 A dynamo and a bulb was added to the SP5 base plate to create the SP5D. A drive band ran from the flywheel to the pulley on the dynamo to power the bulb. As it turned out, this proved ideal for the educational trade as it clearly demonstrated the conversion of steam power to electricity.

Mamod SP5D

2001 The Le Mans racing car was introduced to the line-up of successful Mamods. This model was a twin cylinder racing car and it was hoped that its competitive racing branding would catch on. A red coloured remote control version was also introduced in 2001. However, the manufacture of the RC engine was suspended in 2010.

Mamod Le man Racer

Mamod Le Man RC Racer

2002 After the final movement of Mamod stock and tooling to Birmingham, the previous owners declared that all stock and tooling of the Mamod locomotive range had gone missing. However, a small number of kit locomotives were made in the Birmingham factory from limited stock. It was decided to start from scratch and make more sophisticated models.

Instead of fuel tablets, a new gas burner was developed. The boilers of models were silver soldered. The other first, was with the engines itself to start using a double acting slide valve with slip eccentric. This type of engine was much more powerful than the original Mamod engines. Reheating coils were also added to later models.

Mamod Gas Burner

2003 Rolling stock was introduced as well as the 'O' gauge track.

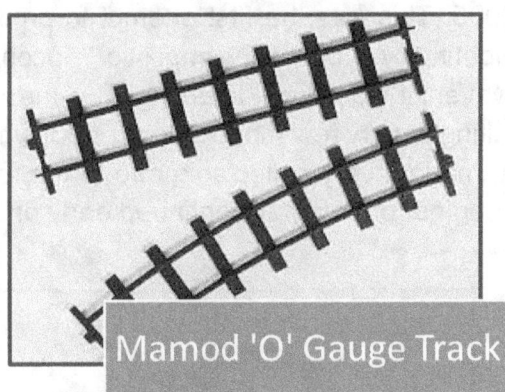
Mamod 'O' Gauge Track

2004 The William locomotive was introduced. Being a very similar model to the Golden Jubilee locomotive, the colour scheme was green with black running towards gold wheels and had the Mamod logo on the sides. It was later discontinued in 2011.

William Locomotive

Next page 2005-09

2005 The SP6 stationary engine was introduced. This was the first stationary engine to use the highly successful double acting slide valve engine with slip eccentric. It was the most powerful engine in the stationary range until the recent introduction of the beam engine. This model ran on fuel tablets

Mamod SP6

Mamod Centurion

Mamod Challenger

Mamod Showman's Special

As well as this, Harry the Rocket was introduced in 2005. This was the first attempt to produce a competitively priced simplistic locomotive. However, it was considered ugly by the public, resulting in very few models being sold, which, in turn, made it very collectable due to the rarity of the engine for being discontinued early on.

Mamod Harry The Rocket

2006 Following the successful launch of the SP6, the Centurion, Challenger and Showman's Special were introduced in February 2006 which all featured the successful double acting slide valve engine with slip eccentric. The boilers were silver soldered with reheating coils at the rear end. All three of these engines had canopies and can be converted to gas by using the scuttle conversion unit.

2007 Unfortunately, by 2007, many model outlets had been closed down (possibly due to the recession) and the few remaining businesses could not afford to stock the large range of Mamod models. The customers of Mamod that were increasing their orders were the ones that were selling on the internet. Therefore, it was decided to relaunch the website and offer direct selling to the public. This helped Mamod to weather the recession and also encouraged many people to visit the factory.

2008-2009 The locomotives Mark 1 and 2 were introduced. These engines looked similar to the original Mamod locomotives, but with greater improvements. Mark 1 locomotives featured improvements to the original specification such as:

- Larger improved boiler
- Safety valve release pressure 40psi
- Butane/Propane gas fired (external)
- Cab sighted steam regulator
- Oscillating double action cylinders with glands
- Improved forward and reverse lever
- Improved wheels and axles
- Silver soldered boiler construction with re-heat tube

Next page 2010 -14

The Mark 2 locomotives featured improvements such as:

- Cab located in-line lubricator
- Smoking chimney
- Brass chimney cowl
- Redesigned chimney
- Brass window spectacles

Mamod Mark 2

Mamod Mark 1

The Mamod SP7 was introduced which was close to identical to the SP6 but had twin cylinders and a pressure gauge fitted to the boiler.

2010 The Brunel locomotive was introduced. This was a 'decoration' type engine which was totally different to previous Mamod locomotives. This was the first Mamod engine ever to have a vertical boiler. Originally, copper was used to make the boiler. However, because of leaks and being prone to dents, the material was changed to brass. It was first issued with fixed wheels either 'O' or '1' gauge and was later modified to be dual gauged by adjusting the centres with an Allen key. Copper pipe was supplied for those that who wanted to direct the steam from the valve chest to the chimney.

Mamod Brunel Vertical Engine

2012 The Diamond Jubilee Saddle Tank was introduced. This was a locomotive that celebrated the Royal Diamond Jubilee of HM Queen Elizabeth II. The engine came in a ruby red rich colour, had brass spectacles, a brass chimney cowl, brass numbered limited edition plaque mounted on the roof, saddles tanks with special decals, came with steam oil and had a gas adaptor. It was a locomotive built on the successful Mark 2 engine chassis.

Mamod Diamond Jubilee Saddle Tank

2013 The SP8 Beam Stationary engine was introduced. This is the latest model to come from Mamod (upon writing this July 2014). The silver soldered model can be run on gas or fuel tablets and uses the tried and trusted double acting slide valve cylinder with slip eccentric (found on the SP6 and other valve engines). It also has an efficient regulator which allows the engine to run on high and low speed and is the most powerful engine by Mamod to date.

Mamod SP8 James Watt Beam Engine

2014 Mamod reintroduces the Marine engine.

Advertisement:

Order on-line direct
www.mamod.co.uk

Made in England since 1937

Brunel
Vertical Boiler Engine

Mamod's First Geared Engine
Silver soldered boiler • Re-heating coil • Slide valve piston/cylinder
Glass water level • Pressure gauge • Ceramic burner • Butane/propane gas tank
Safety valve rated to 40psi • New improved sight glass • Spare port
Internally framed wheels - regaugeable to '0' or '1' gauge

Specification:
175 x 250 x 115mm
Gross weight: 1895g

Mamod Limited
Unit 1A Summit Crescent Industrial Estate Smethwick Warley West Midlands B66 1BT
T: 0121 500 6433 Fax: 0121 500 6309 E: accounts@mamod.co.uk www.mamod.co.uk

The Range Of Mamod Models

Mamod have created a wide range of engines over the years to accommodate every customer's desire. Generally, the models can be grouped into two main categories: stationary and mobile. Here is a brief look at all of the different types of engines Mamod have made over the years.

This is only a brief look at the engines Mamod have produced over the years. For a more in depth look, I would advise you to pop onto the internet to www.mikes-steam-engines.co.uk.

Although the models Mamod have made over the years are mentioned in the 'History Of Mamods' section, here is a summary of the main engines Mamod have made and how they differentiate from each other.

Stationary Engines

MM1. This engine had the piston on top of the boiler. It is still the smallest engine Mamod has ever produced to date. As you can see from the pictures, a solid flywheel on a Minor 1 makes clear it is much older as well as the decal being different too.

MM2. This engine is exactly like the MM1 but slightly larger in dimensions. Think of this engine as the big brother to the MM1. The main differences between the MM2 to MM1 is that the boiler, firebox and flywheel are all slightly larger.

SE1. The SE1 has a larger boiler than the Mamod Minor (MM) range and has the piston, cylinder, crankshaft and flywheel assembly planted onto the same base as the engine via a green engine bracket. The boiler now also has a water level plug which was later converted to a sight glass for health and safety reasons.

SE2. The SE2 engine is exactly the same as the SE1. The only differences or modifications are that the SE2 has is a whistle on the boiler and a regulator or a forward/reverse lever to control the level of steam into the cylinder (therefore, controlling the speed at which the engine runs with the forward/reverse lever also controlling the direction the engines runs in).

SE3. The SE3 is much larger than the SE1/2 and has two pistons connected to a larger boiler. If you find the joints are hard soldered and there is no lever/regulator gear, you may find you have a very rare George and Griffin engine which was designed for use in schools. These engines were very powerful as the singular flywheel was connected to two pistons.

SE4. This is a really rare engine which dates to the late 1940's. The engine has gearing, two pistons and an extra-large boiler. Unfortunately, most of the SE4s produced were either sold in extremely small quantities or were prototypes. The Hobbies SE4s were all hand built by Geoffrey Malins himself. Once Mamod was created, two prototype Mamod SE4s were produced (far left picture). This means that if you find you have an engine that looks like the model below, the chances are that it is a Hobbies SE4 which is still extremely rare.

SP1. In 1979, the SE range was replaced by the SP range. The SP1 went to replace the MM1 with the same size boiler but different styling. A black chimney is a feature of the SP range which does not connect to the piston's exhaust gases (so you do not get to see steam coming out of it). The SP1 boiler is surrounded by a black firebox and introduced a new colour for the engine bracket: blue. As well as this, the whole SP range has vaporising or solid fuel burners instead of wick burners.

SP2. The SP2 has a similar design to the SP1 with the only differences being that the boiler is slightly larger and instead of having a firebox housing around the boiler, it had been replaced with a chrome housing. The SP2 has an adaption with a dynamo to make it the SP2D.

SP3. The SP3 was a Meccano built engine which featured a grey base, chrome surrounding the boiler and a sight glass. The engine had a similar layout to the ME range but was built as a stationary engine to be used on land and used solid fuel. Mamod built this engine after their contract with Meccano expired. It was a hit since owners were easily able to customize the engine with Meccano parts.

SP4. The SP4 was a larger version of the SP2 and featured a larger base where the piston/cylinder assembly was mounted to the side of the engine on a grey platform. The engine now has piping from the cylinder to the chimney so when steamed up, the exhaust gases come out of the chimney.

SP5. The SP5 features a larger boiler than the SP4 and has two pistons. An adaptation of the SP5 was created called the SP5 + Dynamo (far right picture) which has a dynamo and a bulb connected to the flywheel to demonstrate the change from steam power to electricity.

SP6. The SP6 is Mamod's most powerful engine to date (along with the SP7 and SP8). It has a double acting slide valve cylinder and continues the colour theme of the SP range. As well as this, there is a valve to control the steam flow at the top of the boiler. Mamod made a green special edition base for Forest Classics back in 2006.

SP7. The SP7 is very similar to the SP6 model. The main difference is that the SP7 has two double acting slide valve cylinders providing more power to the flywheel. As well as these, there is now a pressure gauge on the top of the boiler.

SP8. The largest and most complicated of the SP range, the SP8 has the same double acting slide valve cylinder and includes the James Watt beam which has a beam connected to the cylinder assembly to the flywheel. This is one the most powerful engines Mamod has ever produced to date (July 2014).

ME1/2/3. Mamod did produce some Marine engines which featured a steam engine surrounded by a chrome housing on a narrow red base (ME3 to the left, ME2 in the middle and ME1 to the right). These engines did not sell well.

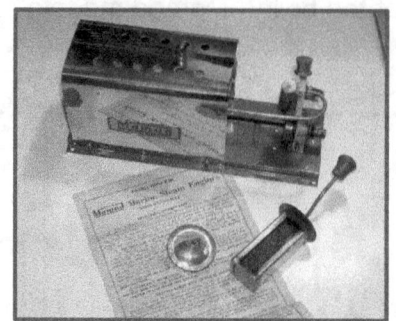

Mobile Engines

TE1a. Mamod's TE1 engine has a flywheel connected to the rear left wheel via a drive band so the engine can move. The TE1a steam engine is a famous representation of the typical tractor steam engine. If the engine does not have a lever but a regulator by the pipe's entrance to the boiler to control its speed, it is a rarer and older engine known as the TE1 which dates to around 1966 or earlier. The TE1/a came with a canopy to cover the engine.

SR1a. This is the Steam Roller which is similar to the TE1a Traction engine but has two rollers at the front instead of wheels. Again, if it has a forward/reverse lever, it is a later SR1a model. If it has a regulator by the pipe's entrance to the boiler, it is a much older and rarer SR1 engine. The SR1 was the first engine by Mamod to have a sprayed boiler in old Apple green.

SW1. The steam wagon is similar to the TE1a traction engine except for the fact that the chassis extends backwards to accommodate the wagon body kit. It first came in green and blue and is currently still going in blue and brown on Mamod's official website.

SA1. This is Mamod's steam roadster which is commonly featured in white. Its design is based around the American Mercer Raceabout from the early 1920's. It has four tyre wheels and leather seats with the forward/reverse lever to control the speed and direction to the left of the driver's seat. The roadster comes in burgundy or green (also known as the Brookland Tourer).

SA1L. This is the limousine which is based around the Mamod SA1 but has extra seats (two more) with a roof over the top. This engine either comes in silver or burgundy.

Delivery Van. A new edition to the Mamod mobile range, the delivery van is based around the SA1 and can be differentiated by these models as to being a 1920s styled delivery van with the Mamod logo on the side of the van. This engine either comes in blue, green or black.

FE1. Mamod produced a fire engine which comes in bright red and, just like the SA1L and delivery van, is based on the SA1. Unfortunately, the hose is not operational but merely there for aesthetics.

London Bus. The London bus by Mamod comes in bright red or green and is a double-decker at that too. If you manage to have one of these, you are very lucky indeed as personally it's my favourite Mamod as I see it as one of the best looking Mamod engines out there.

TWK1. This was the first ever kit Mamod offered where the consumer had to build the engine. The kit was of a TE1a traction engine, without the canopy roof, in matt black and yellow and had a lumber wagon to pull behind it. It used parts from other engines. For example, the front wheels were from the SA1 while the flywheel was from the SW1.

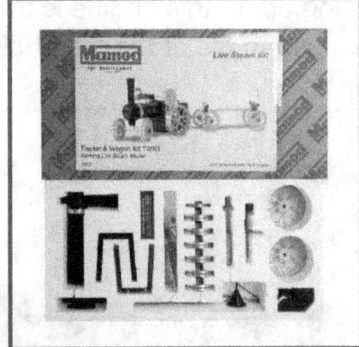

RS1/2. Mamod produced a railway set where the train had an olive green (or blue) coat and two pistons on either side of the boiler. This was quite a powerful engine (RS1 on left with the RS2 on the right).

Vertical Brunel. A new addition to the Mamod's trains was the vertical Brunel engine. It was Mamod's first geared engine, which had a reheated coil and was hard soldered like the SE3 Griffin and George engines.

As you can see, Mamod have made quite a lot of engines over the years and due to improvements in manufacturing and technology, the British toy manufacturer has been able to make more complex and powerful engines such as the SP6/7/8 engines.

Although the below engine is not actually made by Mamod, I thought I would include it anyway since it is a beautifully designed engine that I would love to own one day (when I can get the funds!). With toy steam engines, there are two main competitors: Mamod and the German toy manufacturer Wilesco. In general, Wilesco steam engines are more complicated which can sometimes work to both their favour and against them (simplicity often wins with toys sometimes).

The below engine is the Wilesco D305 which is Wilesco's take of the steam fire engine. What is so amazing about this engine is that the hoses are operational, meaning you can either use the power of the engine to make the fire engine move or to power the pump allowing water to spray out of the hose.

Advertisement:

Order on-line direct
www.mamod.co.uk

Twin Cylinder Working Steam Engine
C88

SP5D

WITH DYNAMO

Generator and light included. Pressure gauge available as an extra. Also available without dynamo. Forward running only.

Specification:
Pack Dimensions: 250 x 210 x 195mm
Gross weight: 1500g

Mamod Limited
Unit 1A Summit Crescent Industrial Estate Smethwick Warley West Midlands B66 1BT
T: 0121 500 6433 Fax: 0121 500 6309 E: accounts@mamod.co.uk www.mamod.co.uk

Steaming Your Engine Up

Along with restoring steam engines in this book, it is important to know how to correctly steam up your Mamod steam engine so that you can get the most out of and prolong the life of your steam engine (by the way, the above picture looks as if the piston rod is bent, but that is only because the speed of the piston moving makes the piston rod look as if it is bent – it is an optical illusion).

It is especially important to remember that some of the engines you attempt to run will be tens and tens of years old and were built in a time when toy legislation was a bit more forgiving than the modern era. For this reason, always take every safety precaution possible when steaming your engine up (especially for the first time). Expect for bad things to happen so you can prepare for the worst scenarios (such as a part flying off at high speed). It is advisable to wear thick gloves and eye protection and have a supply of water/fire blanket handy.

As a steam engine enthusiast, great care should be taken when steaming up. This is because you are not working with common 'steam' such as the steam from a boiling kettle. You are dealing with steam under pressure which has even more energy and can reach much higher temperatures than 100 degrees Celsius. The water inside a boiler, when being heated for steaming up, will boil creating extremely hot steam under pressure. This steam needs to contain lots of energy. After all, it has to have the power to push a cylinder out and rotate everything that is connected to the crankshaft.

If you choose to steam your engine up without taking the necessary factors into consideration, you are putting yourself and everyone else around you at risk. For example, I was once gifted a TE1a traction engine/wagon kit to build. I had built it correctly. However, the safety plug was only loosely screwed on and under the pressure from the boiler, flew off at lightning speed. Luckily for those around me, it did not hit anyone or anything (although it did land around 10 metres away from the engine). Steam engines are great pieces of kit but they are dangerous pieces of kit if mistreated. That is why you should always be extra careful when dealing with a live steam engine.

Preparations and Safety

When it comes to steaming up your engine, there are a few things you will need to have with you to ensure you steam up in absolute safety:

- Mamod steam engine, of course.
- Fuel for the engine - This can be one of many such as solid fuel tablets, methylated spirit, gel or gas.).
- Oil - This will be used to lubricate the piston/cylinder assembly to reduce friction and the possibility of wear on the engine.
- Gloves to protect yourself against any heat from the engine.
- A box of extra-long matches.
- A bucket of water/sand and fire blanket - You must always remember that Mamod steam engines are still dangerous pieces of equipment if treated without care. Therefore, it is always essential to have a bucket of water/sand and a fire blanket handy in the eventuality of the worst situation of the engine or something near the engine going up in flames.
- A kettle.

If You Use Methylated Spirit, Have a Bucket of Sand and a Fire Blanket Instead of Water

If you find yourself in a situation where you will need to extinguish a fire, **do not** ever extinguish a fire that has been caused by methylated spirit with water. This is because the water will only spread the fire making the whole situation become more dangerous and uncontrollable. This is why for methylated spirit fuelled engines only have a bucket of sand or, better yet, a fire blanket that you can smother over the engine in the worst case.

Methylated spirit is a dangerous chemical to say the least. It is extremely flammable, spreads easy and has a colourless flame. Therefore, if you choose to use this liquid fuel, make sure you are filling the spirit burner on a flat surface and take your time! The most dangerous point involves manoeuvring the burner into the firebox once the spirit burner is lit. Be extremely careful not to let the burner spill methylated spirit as it will burn you and damage your engine. This is a reason to never overfill when using liquid fuel.

The Engine Will Get Hot

Although this might seem obvious, there are still people that will try to touch certain areas of the engine while it is running, such as the cylinder, safety valve or the whistle. If ever you want to touch your engine while it is hot, make sure you wear thick gloves.

It is also common for Mamod engines to spit oil and water when steaming up. This occurs most when the boiler is at its maximum pressure at which point the safety valve will be releasing steam from the boiler. If you choose to open the whistle at this point, you should not be surprised that hot steam and water from the boiler will come out of the boiler, potentially hitting your hand and burning it. This can also happen in the cylinder/piston area.

Refill the Boiler Every Time the Fuel Runs Out

It is extremely dangerous to run an engine on two sets of fuel while keeping the same water in the boiler. As you run the engine, the water level in the boiler will naturally decrease to a point when the heat from the fuel will not be going to the water (since there is not much left in there) but the surroundings. This can cause areas of the boiler to melt starting with the washers and solder joints. For this reason, every time your fuel has burnt out, hold down the whistle to let all of the steam out and then refill the boiler.

What you must always remember is that steam engines are always going to be dangerous pieces of equipment to run: especially the old ones. This is because the regulations and laws for steam engines tens of years ago were a bit more forgiving than they are now, and put more responsibility in the user of the steam engine than the steam engine itself (this is a reason why most of Mamod's engines are now hard soldered).

You are not just dealing with the type of steam you get from boiling a kettle. The steam used in steam engines is under pressure, which means that it holds much more energy than 'kettle steam' and reaches much higher temperatures beyond 100 degrees Celsius. I have had only one encounter of something blowing off of an engine and it was quite a shock at the time. It seemed like a jet of steam shot out from where the safety valve had been. The safety could not have been fully engaged with the thread. If I was close to the engine, it would have seriously burned me. So, please remember how dangerous these engines can be and always take safety seriously.

Safety When Restoring Your Engine

Not only can it be dangerous when steaming up an engine, it can be just as dangerous in some of the processes used in this book to restore your engine. For this reason, here are some safety precautions you should take when restoring your engine:

- Wear safety glasses when using any power tool. Bits of polish, metal and dirt can fly off and hit your eyes.

- Do not wear any loose clothing (and if you have long hair, tie it up). Tools such as a drill could catch in any loose clothing and hair.

- Be careful when using white spirit. White spirit is both irritant to skin and dangerous to the environment. Therefore, if you get white spirit on your hands and skin, use a good hand cleaner to remove the white spirit (if your skin becomes dry, use a moisturiser after cleaning to hydrate the skin).

- Wear protective gloves when in contact with heat. You should wear gloves to protect you from the heat of the engine while steaming up, a solder iron or torch.

- Wear a suitable mask when spraying and removing paint. Aerosols and old paint on some Mamod engines have solvents and lead in them which is harmful to humans. Always try to spray paint and remove old paint from parts in a well-ventilated area to disperse the harmful chemicals better.

General Tips When Steaming Up

Of course, with any engine, it is easy to get it running. For a car, this involves turning a key and waiting for the starter motor to get the engine to turn. For a Mamod engine, this involves putting water into the boiler and fuel under the boiler and then waiting for the internal pressure of the boiler to rise and to then give the flywheel a quick flick to get the engine going.

However, all engines are subject to services. A car should get a service around every 10,000 miles. This makes the point that Mamod steam engines still need to be serviced in order to achieve the best steam ups. Here are some tips you should do every so often to make sure your engine runs to its optimum. Remember that some of the engines you will be steaming up will be many years old and will be in fragile states in need of TLC:

- Flush out the boiler with limescale remover. The constant contact with water causes limescale to form in the boiler, resulting in your boiler becoming filled with impurities, thus hindering the engine's performance. It is worth putting a tiny amount (such as a teaspoon) of limescale remover mixed with warm water into the boiler, shake it a little and then flush the system out. Once you have done this, put cold water into the boiler and flush the water out again. I found an effective way to tell how much 'rubbish' is in the boiler is by tilting the whole engine up a bit and lightly shake. Looking into the sight glass (if your engine has one) will allow you see how much 'rubbish' is in the water.

You might also want to try an old technique of running the engine with a little vinegar in the boiler. I have not tried this personally. However, from looking on the internet, it seems an effective way to clean the deposits out of a boiler: even if it does smell when steaming up! Do not hold my word for it though!

- Replace fibre/rubber washers. Over time, the constant pressure exerted on washers will cause them to slowly deteriorate until they are not gas tight anymore. Take off parts such as the safety valve, water level plug and whistle and see how the washers are. If they look damaged, you might want to replace them since a damaged washer will lose your engine steam thereby decreasing performance.

- Replace O-rings. If the piping to your boiler looks like the image to your right, then your engine has an O-ring which, like washers, can deteriorate over time. Have a look at P99 for a more in depth look at the O-ring on Mamod engines.

- Lightly clean the engine. After a steam-up, if you do not clean your engine, you are leaving the engine covered in oil and water. Over time, this will cause your engine to rust. Therefore, clean your engine every so often with an old rag making sure it is as dry as possible before storing - always pour the water out of the boiler after steaming up.

- Clean the burner tray. This tip is only with solid and gel fuel burner trays. Once solid fuel or gel has burnt out, it will leave a residue at the bottom of the burner tray, which, if you do not clean it, will build up and become quite impossible to get off. Therefore, it is always best to clean the area where the fuel sits in the burner tray so you can maintain the condition of it (image to the right shows what will happen if you do not clean it)!

Instructions

The first step to steaming up your engine is to make sure you have safely prepared the area you are going to steam up in. This involves making sure your engine is in a well-ventilated area and away from anything that may be flammable. As well as this, it is important to always have a bucket of water around a metre away in case anything does happen (or a fire blanket or a bucket of sand if you are using methylated spirit).

As well as this, you need to check that your engine is in a perfect working order state so it is *possible* to steam the engine up. Before doing anything else, make sure you check the following areas on your engine:

- The safety valve. You need to make sure the spring in the safety valve works so it is capable of releasing steam from the boiler preventing the pressure in the boiler from getting too high.

- Piston/cylinder assembly. The piston must be able to freely move in and out of the cylinder while the cylinder rotates freely on the engine bracket.

Once the area you're steaming up in is prepared, it is time to fill up the boiler with water. The reason for having a kettle is because if you do not use a kettle to preheat the water before it goes into the boiler, you will have to heat the cold water up inside the boiler using fuel on the burner tray. This method is more expensive (as you are using Mamod fuel) and will take much longer than using a kettle to heat the water up. I am sure most people have a kettle so this should not be a problem.

When tightening anything with threads onto an engine, make sure it is only finger tight. Any tighter (such as tightening with pliers) will make it hard to take either the water level plug or safety valve out which can be very dangerous and also damage the washers.

While the kettle is boiling, it is advisable to set up the remaining elements for steaming your engine, since as soon as you put the water from the kettle into the boiler, the water will start losing heat. Therefore, you will need to get the burner tray with lit fuel underneath the boiler as soon as possible to obtain the longest steam up time. For this reason, put the fuel onto the fuel tray and have matches at the ready to light the fuel. Just remember to not light the fuel until the boiler is full, though.

Once the kettle has boiled, it is time to place the water from the kettle into the boiler. It is extremely advisable to buy a Mamod official or unofficial funnel to guide the water into the safety valve hole into the boiler - without a funnel, the water will go all over the boiler and, in the long-term, will lead to lime scale marks and rusting. If your engine does have a water level plug, take this out. As soon as water starts coming out of the water level plug hole, stop filling the boiler up with water; as there is now enough in the boiler to steam up on. Clean any spilled water with kitchen cloth and screw back in the safety valve and water level plug finger tight. If your engine has a sight glass, fill the engine up to the lip at the top of the sight glass.

At this moment in time, you should have boiling water in your steam engine's boiler and will now need to put the fuel for the engine underneath the boiler to continually heat the water up to create steam pressure. Carefully, light the fuel for your engine in its tray and place the tray underneath the boiler in the firebox. When it comes to lighting the fuel, you need to be careful depending what fuel you use:

- Solid fuel - Solid fuel is usually quite difficult to light on a burner tray. The best way to light the tablets of solid fuel on a burner tray is to tilt the tray and light one of the tablets. Once this tablet is burning away, it will start to light the other tablet if you continue titling it so the flames from the first tablet engulf the second tablet. When it comes to buying the fuel for your engine, it does not really matter if you buy Mamod official or unofficial tablets: they all work the same and last roughly the same as well.

- Methylated spirit - This is the most dangerous of all the fuels you can use on your Mamod engine. However, methylated spirit is much cheaper to buy. The dangerous element to methylated spirit is that when the liquid burns, the flame is completely invisible. Therefore, it is difficult to distinguish when the liquid is lit or not. As well as this, if any of the methylated spirit is spilt and is alight, it will continue to burn. To counteract these hazards, don't overfill your spirit burner: once you see the methylated spirit start to come through the mesh that is your sign to stop filling the spirit. When it comes to lighting the methylated spirit, use a match and touch the flamed end of the match with the mesh on the spirit burner. Once you have done this, give it around 5 seconds and place your hand 20-30cm above the burner. If you feel heat coming off the burner, it is lit and you can put the burner into the firebox. As well as this, if you look at the burner side on when lit, you will see air distortion from the heat of the methylated spirit burning (another sign you can use to tell the spirit burner is alight). Remember that to put a methylated spirit out you will need either sand or a fire blanket as water will simply spread the fire.

- Gel - Personally, this is my favourite fuel as it is safe, cheap and effective at steaming up engines. Gel fuel is basically a gel that is flammable. You can use gel fuel with the traditional solid fuel burner. When lit, it will produce a visible flame. The only downside to gel as fuel is the residue it leaves on the burner tray which, over time, you will have to clean every so often to keep your burner tray in good condition (although I tend to dedicate one tray to gel fuel and just let the residue build up).

- Gas - Gas is the most modern of the fuel used to fuel Mamod steam engines. The only real downside to gas is the cost of buying the gas burner in the first place which ranges from £70 upwards. However, in the long term, the fuel is much cheaper and it is a much cleaner burn for your engine too (no soot will develop underneath the boiler). It is a good investment if you steam up frequently since in the long term it will work out cheaper and better for your engine.

Once you have lit the fuel on your burner tray, carefully place the burner tray into the firebox. You will have to wait at least a few minutes for the temperature of the water to get high enough to steam up the engine. There are a few ways of telling when your engine is ready to steam up:

- Every so often, quickly tap on the whistle (if you have one). If it sounds like air coming out, you need to wait a bit longer. If the whistle makes a high pitch sound, you are most likely ready to run the engine.

- Watch the safety valve. The safety valve on all Mamod engines is designed to release steam if the pressure in the boiler becomes too high. Therefore, if you see that steam is coming from the safety valve, it is time to run the engine.

- Every so often, attempt to get the engine running by flicking the flywheel to get the piston/cylinder assembly to rotate. If nothing happens, the engine still needs time for steam pressure to build.

By now, you should be at a stage when the engine is ready to run. For most Mamods, you will need to help the engine start by rotating the flywheel to give the engine a bit of momentum. Following this, you should have your engine running!

In most cases, the fuel runs out before the water in the boiler does. When the fuel does burn out, always top up the water in the boiler if you want to steam up your engine again. If you decide to do this, remember to let all of the steam out of the engine before you take the safety valve and the water level plug off - if you don't let the steam out they will fly off potentially hurting someone close by. The way you can let the steam out of the engine is to hold the whistle down until no more steam comes out of the whistle.

Make sure to pour every drop of water out of the boiler and clean all of the engine once you have finished so that it does not rust the engine or damage it in the long term. Run the flywheel with your finger so all of the condensed water comes out of the piston/cylinder assembly and dry the engine ready for storage.

Advertisement:

Made in England since 1937

Order on-line direct
www.mamod.co.uk

Brooklands Tourer
1319BT
Working Steam Model
C142BT

A magnificent working model of a steam car that captures the elegance of a bygone age.
Available in green, burgundy or cream.
Also available in kit form.

Specification:
Pack Dimensions: 450 x 150 x 210mm
Gross Weight 2450g.

Mamod Limited
Unit 1A Summit Crescent Industrial Estate Smethwick Warley West Midlands B66 1BT
T: 0121 500 6433 Fax: 0121 500 6309 E: accounts@mamod.co.uk www.mamod.co.uk

Buying Your Engine

Possibly the most important aspect to the whole restoration process, buying the wrong engine will result in you putting far too many hours and money into the restoration. In this section, I will highlight that there are *certain* types of engines you should buy to make sure your restoration is possible, relatively cheap and enjoyable. As well as this, I will also highlight the best websites where you can buy your restoration engines and parts on page 101.

What Do You Want?

The first question you need to ask yourself before you buy your engine is what do you want? Depending on your situation determines the type of engine you will want to buy:

- A tatty looking runner   - A non-runner

However, before you decide this, you need to select what model engine you are going to restore.

What Model Engine to Restore?

Choosing the model engine is the first step to your restoration. To make it easy for you, I have categorised all of Mamod's steam engines from over the years into either easy, slightly difficult or difficult restorations. This should help you choose, based on your experience in restoring engines, to elect the right engine for you.

Easy Restorations

These engines can be considered the easiest restorations to do out of all the Mamod engines. This is because they have the fewest parts, are the cheapest and take the least amount of time to restore.
(Continue onto the next page)

(Prices of engines are estimates from July 2014 and will generally increase over time as Mamod engines become more collectable and older).

Minor Range

For anyone wanting to do their first restoration, I would recommend starting out on these engines. The Minor range are the simplest engines by Mamod because they have so few parts.
The other great thing about these engines are that they are ridiculously cheap - the cheapest of all the engines. You could pick yourself one up for £20-£30 depending on the condition of the engine.

The Minor range has two types of engines: the MM1 and MM2. If you are unsure which to buy, remember that both are near enough the same: the only real difference is that the MM2 has a slightly larger boiler than the MM1.

SE Range

The SE range are all simple engines because they only have four painted parts and the rest can be metal polished.
The other great thing about these engines are that, like the Minor range, they are also quite cheap. You could pick yourself up:

- SE1 for around £35.
- SE2 for around £45.
- SE3 for around £60-£70.

If you do not know what SE engine to choose, think of the SE1 as the easiest, rising in difficulty with the SE2 and then the SE3. However, since these engines are all similar in the parts they have, paintworks and layouts, it will not matter too much on what engine you choose - I would advise you to opt for the engine you like the most (for me, it's the SE3).

SP Range

The SP range was the replacement to the SE range. Therefore, they share similar characteristics with the SE range which means that they are just as easy to restore.
The only problem with the SP range is their pricing. Mamod currently (2014) are still producing the SP range. For this reason, they hold their value slightly better than the SE range. Therefore, you will be looking at a cost, for any SP engine, of at least £40 and up.

As well as this, the parts for the SP range are also slightly more expensive. The chances are that the chrome around the boiler on an SP2/4 has rusted away so that the chrome layer has been removed revealing a dull grey steel colour. If you want to buy replacement chrome, it will cost you around 2-3 times more than it does to replace the chrome on an SE model.

If you manage to buy an SP1, try not to restore the paintwork since these engines are rare and are worth more in their original condition since they are old engines.

Slightly Difficult Restorations

The easiest restorations are the stationary engines because they have the minimum amount of parts for them to run. The following engines are considered slightly more difficult to restore for the reasons that they have more parts. Therefore, there is a wider range of parts to restore, together with more parts that could potentially fail.

TE1/a Traction Engine

The TE1/a traction engine is slightly harder to restore since the torque from the flywheel is transferred to the rear wheel of the engine via a drive band. You are not just trying to make the flywheel rotate like you would be doing on a stationary engine. You are now trying to make the whole engine move forward (or backward).

Along with the added parts such as the wheels and front axle, the paint job on the TE1/a engines is more complex. The traction engines feature heat resistant paint on the boiler. However, it is difficult trying to find the correct colour paint that Mamod used, whilst making sure it is not matt, but is also heat resistant. The prices of such paint can be quite high...

To get a better understanding of how to restore a traction engine (or steam roller), have a look at the section 'Restoring a Mobile Engine' (see pages 87-94).

It has to be said the satisfaction of restoring a traction engine cannot be beaten since this is the essence of owning a steam engine. You could expect to pick up a traction engine for £40 and up.

SR1/a Steam Roller

The steam roller is equally as difficult to restore as the traction engine since it is almost identical except for the wheels at the front are replaced with rollers. You can find steam rollers selling for similar prices to the traction engine at £40 upwards.

What I would say about the traction engine and steam roller is that because of the difficulty in trying to replicate its original paint scheme, many restorations end up creating a unique paint scheme as it is usually easier and makes the engine look better too.

Additionally, restorers like to add touches of customization to their engines. For example, the brass coloured traction engine on the previous page actually started out with a green boiler. However, it was easier for me to strip the paint and turn it into a brass boiler traction engine (looking nicer than the green boiler traction engines, in my opinion). As well as this, the steam roller I restored had a black boiler instead, which actually made it look a quite distinguishing (again, another customized paintwork). This reinforces that is it much easier to add a touch of customization to these engines than the stationary which is something restorers should be encouraged to embrace.

Difficult Restorations

The reason why all these engines are considered difficult is because:

- They have more parts to restore than the slightly difficult restoration engines.
- The structure of them is different (such as the wagon with has a more complex chassis and bodywork).
- The paint jobs for these engines are more complicated.
- The cost to restore them are much higher.
- The price of the engines are much higher.

All of these engines, in my opinion, are considered difficult restorations compared to the engines already mentioned in the 'Easy' and 'Slightly Difficult' restoration engines:

- All Mamod steam trains.
- Mamod SW1 Wagon.
- Mamod SE4 engines (however, if you do get one of these, don't restore it! They are extremely rare and worth much more in their original condition).
- Mamod Steam Roadsters.
- Mamod Fire Engine, London Bus and Delivery Van.
- Any other Mamod engine I have not mentioned.

The point on the SE4 makes clear that for extremely old engines (which you will be able to identify by looking at the 'History of Mamod' on page 8 onwards and 'The Range of Models' on page 21 onwards) under no circumstances do you restore them! Extremely old and rare engines are worth much more in their original conditions. The only thing I would advise is to clean the engine with a cloth and cleaning products - that is all.

I do feel it would not be right to restore some of the really old engines. They have survived tens of years and some have even lived through World War II! It would be a shame to restore these engines as, in part, you are removing a chunk of their history: their condition tells us a story in itself. The only reason to restore a really old engine is if it is in an absolute dire state such as the one I restored, as demonstrated on the next page.

⬆ The engine was a complete non-runner.

Hopefully, you now have an idea of what model engine you want to restore. Now that you have decided, the next step is to choose either a tatty looking runner or a complete non-runner.

If You Want to Restore a Tatty Looking Runner...

Restoring a tatty looking runner is the first type of restoration inexperienced restorers should opt for. You need to understand the principles of how a steam engine works and is put together before you attempt a non-runner.

In general, a tatty looking runner restoration involves completely taking apart the engine, restoring each individual part and then putting the whole engine back together. The great thing about this type of restoration is the final result and the satisfaction that you have just brought back a Mamod steam engine to its former glory.

When it comes to choosing a tatty looking runner, you need to take into account a few things first.

What is the paint work like?

On some engines, the paintwork could be completely fine with just some dirt on it. Ideally, you want to leave as much of the paintwork as original as possible. This will help increase its overall value. As well as this, the coat Mamod put on their engines will always be the best so try to leave as much of it original. The only problem is that with a better coat on the engine, the more expensive it will be to purchase.

 This engine looks to be in a bad state because it is covered in layers of dust and dirt. However, in actual fact, the paintwork on it is still in good condition.

If you choose an engine which needs new coats of paint, you need to think about what parts need painting? For example, if you wanted to restore a traction engine or steam roller, you will be wanting the boiler's coat of paint along with the firebox to not need respraying since if they need respraying you will potentially have to separate the components (and the restoration will be that bit more difficult). Parts such as the wheels/flywheels, if in bad condition, do not really matter as these are extremely easy to restore and repaint.

Is there any damage?

Just because the engine is a runner does not necessarily mean the engine is not damaged. You need to examine the photos provided by the seller or website and decide if any parts are damaged or not. By damaged, I mean, cracks, rust, corrosion and dents. If there is damage, is it bad enough that you will have to buy a new part or can it be restored? With damage, it is all about weighing up how aesthetically pleasing you want your engine to look and how much money you want to invest into the restoration of the engine.

How 'tatty' is it?

The last question you need to ask yourself regarding the engine is that of the overall condition of it, which will give you a good idea of how long it will take to restore the parts. For example, if some of the metal parts are covered in rust, it will take longer to restore them since you will have to de-rust the parts first before anything else is addressed. Therefore, when looking at how tatty the engine is, look at how much rust is on the metal parts, how shiny the chrome is (if not shiny, the chrome layer has gone) and how tarnished and sooty the boiler is.

If You Want to Restore a Non-Runner...

Restoring a non-runner is going to be pretty difficult. For this reason, I would only advise an experienced restorer who has already restored a few tatty-looking runners before attempting to restore a non-runner.

Since you should know what model engine you want to restore as a non-runner from 'What Model Engine To Restore?' (see page 40), the next step is to look at what engines are available to restore as non-runners from websites such as eBay or the internet in general (good search phrases are 'Mamod [model name] spares' and 'Mamod [model name] restore'). Once you have found some non-running Mamod steam engines, you need to assess *why* it is not running. For this, look at the section 'Problems with Starting Your Engine' (see page 98). If there are parts missing, then it is obvious why it will not start. However, even if there are parts missing, that does not automatically mean that the parts that are there are all in fully working order. For a non-running engine, it is extremely important to look at every aspect of the engine to get a sound understanding of why it is not working so that you can get a clear idea of how much it will cost and how best to restore it.

It is important to weigh up the amount of work that is required to restore a non-runner. For example, take the two engines on this page. If you wanted to restore a non-runner, based solely on the pictures to the right, which one would you go for and why? If you wanted an easier restoration, I would go for the traction engine since the boiler is intact and it only seems like there are a few parts missing (engine just looks extremely dirty). The stationary engine is going to be a more expensive and time-consuming restoration since the picture shows that the boiler is a right-off unless the end cap and threaded inserts are replaced (with the engine having missing parts too).

When you look to buy a non-runner, analyse *every* possible detail of the engine, ask yourself what this will add to the restoration and then weigh up whether it is worth the time and money to renovate the engine. Of course, the harder the restoration, the more satisfying the feeling when you see the engine back to its former glory!

I generally prefer to restore non-runners for the simple reason that it is almost like I am bringing back an engine to life that has been neglected for far too long. This is why I disapprove of people who seek engines that are bought in fully working order and are subsequently taken apart to be sold as spares. Although this may make more money, I think it is sacrilege to tear apart something of such historical value. After all, shouldn't we all be buying steam engines to relive the evocative years of steam billowing from chimneys, the smell of fuel burning and hearing the sound of the whistle blowing?

Advertisement:

Made in England since 1937

Order on-line direct
www.mamod.co.uk

Fire Engine Kit
Working Live Steam Model Kit
C137

This kit is slightly more challenging and will build into a
magnificent working model of an early **Edwardian fire truck**.
Finished in typical bright red with brass and chrome trim.

Specification:
Pack Dimension: 585 x 540 x 100mm
Gross weight: 3050g

Mamod Limited
Unit 1A Summit Crescent Industrial Estate Smethwick Warley West Midlands B66 1BT
T: 0121 500 6433 Fax: 0121 500 6309 E: accounts@mamod.co.uk www.mamod.co.uk

Clues to a Damaged Engine

Nobody wants to unintentionally buy a damaged engine and identifying damage on an engine can be difficult to see not least since most of the signs are inconspicuous. However, if there is any fault with any engine, this damage will always leave clues around the engine which will help us, the restorers, get a better idea of the state of the engine. Here are some key areas you need to look out for on an engine if you are looking to buy.

Location of Rust

The location of rust on an engine will give a great idea to how that rust got there. If the whole engine is covered in rust, it is clear it has been stored somewhere exposed to moisture over the years such as a shed or garage.

The most common area for rust to appear is on the firebox of SE and MM engines, the chromebox on mobile engines and the chrome panels on the SP range. This is because it is usually in contact with water. If you find that there are random areas of your engine that have rust, such as just the engine bracket, you need to ask yourself why?

Let's take the example of an SE1 engine in good condition except for the engine bracket which is covered in rust. It is clear that the part has been in contact with water. Therefore, is there a leak in the piston/cylinder assembly which is exposing the engine bracket to water? Does this mean that the piston and cylinder could be damaged? Is there a leak in the pipes to the cylinder? Just from the location of rust, you can get an idea of what may need to be done when restoring the engine.

Another example could be that there is rust on the firebox surrounding the chromebox on an TE1a traction engine. This would be strange to see since if there is rust on the firebox surrounding the chromebox, you would expect to see rust on the chromebox too. The fact there isn't could mean that the owner of the engine decided to replace the chromebox because it was covered in rust. If that is the case, check the rivets to see if they are Mamod fitted rivets (small closed rivets). If not, it is a clear sign the chromebox has been replaced. What else has been changed on the engine? If the chromebox hasn't clearly been replaced, what made the firebox rust? The location of rust on an engine reveals everything.

Untested

Buying an untested engine is going to always be a risk. Unfortunately, some people are not as honest as they should be. If they have said it is untested, they could be telling the truth. They could have also tested the engine and found out there is something wrong with it as it does not run. They could then claim the engine as a 'barn' find and untested since they allegedly do not know much about steam engines. It is a risk that could pay off if you find the right engine. But, it could also leave you with an engine that could require a lot of attention to restore.

Painted Chrome is a Lost Cause

If you are looking to buy an engine that has the chrome parts of it spray painted, you need to ask yourself *why* those parts are painted. The most logical reason for spray painting chrome is because the chrome has been removed due to rust revealing the underlying dull grey steel. Therefore, you are most likely going to have to buy new chrome parts if you want to return the engine to its original condition.

How the Safety Valve/Water Level Plug/Whistle Turns

An area which can become quite a lot of work for a restorer is the area on boilers because the threaded inserts in the boiler are difficult to replace if damaged. When you are buying an engine, you need to carefully examine the threaded inserts on the boiler and even ask the seller questions such as:

Does the Safety Valve, Water Level Plug or Whistle Unscrew?

Mamod made all their engines with threaded inserts in the boiler for the safety valve, water level plug and whistle. Therefore, if the seller informs you that they do not unscrew, it is clear there is a problem with the threads. Either the parts have not been unscrewed in a long while, meaning there is a severe build-up of dirt and rust preventing the unscrewing or the threads, or the parts are damaged. Either way, it means a bit more work for the restorer.

It is also possible for the threaded inserts to turn rather than the safety valve, water level plug or whistle unscrew. If this is the case, the solder joint that holds these threaded inserts in a fixed position in the boiler has failed, requiring the restorer to resolder the threaded inserts back into the boiler.

Does the Safety Valve, Water Level Plug or Whistle Turn Freely?

If the seller makes clear that these parts turn freely on the boiler rather than unscrewing, this is a more serious problem. The fact that it turns freely means that the parts are not engaging with the thread. This means that either the thread in the boiler has eroded down on the screw, or the parts have eroded down. Preferably, you would want the loose part to be the problem since if it is the insert, to repair this, you will have to take the threaded insert out and replace it with a new one which takes time in removal of the old insert (since you will have to take the end cap off to remove the insert) and soldering in the new one.

Are there White Marks by the Safety Valve, Water Level Plug, Whistle or on the Boiler?

White marks on the boiler is an obvious sign that the boiler has been in contact with water for too long as there is a strong possibility this is limescale. Limescale is the term given to the minerals in water, which come out of solution and deposit themselves onto the surface of a material (and is more severe in hard water areas). The great thing about limescale is that it does not damage the engine and can be cleaned with either a bit of emery paper or a cloth and brass polish.

Limescale can clearly be seen on the inserts and chimney of this SE boiler

The main problem is that if there is limescale outside the engine, there will be limescale inside the engine too. This will result in your engine losing power since the limescale will adversely affect boiler efficiency. As well as this, for the really old engines, there might be a slight chance that the piping has furred. This is when the limescale has built up inside the piping, causing the overall area of the pipework to decrease. However, this usually only happens with piping that has been in contact with limescale for a *long* time: the chances that it has happened to a Mamod steam engine is quite remote. However, for the older engines, it may have happened. If your piping has furred, you can either replace the piping completely or try to remove the limescale using limescale remover.

A furred pipe

Dezincification

The worst possible situation you may find yourself in with white marks on your boiler is dezincification (also known as selective leaching). This is when the zinc from the brass alloy (made from copper and zinc) splits away from the copper atoms and leaves the structure, depositing itself on the surface of the now-honeycomb copper (honeycomb as there are holes in the structure from where the zinc once was). Since zinc oxide is white in colour, the white marks 99% of the time will be limescale. However, this doesn't stop the chance that it could be dezincification. The problem is that if your brass parts have undergone dezincification, they will not be gas tight anymore causing your engine to lose performance or not run at all.

If you find that your engine has undergone selective leaching, you will have to replace whatever parts that have been damaged by the corrosion. This makes clear that selective leaching is a serious matter for Mamod engines that will cost restores a lot of time, money and effort.

> The following paragraphs about dezincification are not fact but based on experiences with engines that have had dezincification. Don't take the next paragraphs as gospel truth but bear in mind that the assumptions are based on past experiences with Mamod engines that have had selective leaching.

From talking to associates about dezincification on Mamod engines, it is caused by a number of reasons. The most common reason involves water which is either slightly acidic or alkaline, has too much oxygen and carbon dioxide in it, low in salt content, or has chlorine ions in it. Basically, depending on what type of water has been used with the engine for steam ups over the years depends on the rate of dezincification.

As well as this, one of my associates has found that the engines, all except one, always have the dezincification on the end caps. He found that the engines with dezincification usually tend to show up from the 1950s. Assumptions for this include:

- The quality of brass used on engines in the 1950s was inferior.
- The end caps were made via a different process to the other brass components.
- The end caps came from a different source of stock material.

The above are *just* assumptions to why dezincification occurs based on knowledge from that time. However, the main reason for selective leaching will be neglect from owners. Restorers are unlikely to cause dezincification because we care for these engines. However, there was once a time, though, when these engines were treated as just toys. This means water was left in boilers while being stored and the engines were not maintained properly.

Unfortunately, dezincification on Mamod engines is a very 'woolly' topic because it is a rare occurance. For example, I have never had an experience with dezincification with any of my engines and I have restored some extremely old engines over the years. This doesn't mean you won't experience dezincification, though. All I would say about this is that the older the engine, the more likely there is dezincification which means you may need to replace some of the brass parts.

How the Steam Engine Works

Understanding how a steam engine works is extremely important when it comes to restoring it. A steam engine is an external combustion engine. This means that the flame (which is used as the source of heat energy) is outside the engine. A petrol/diesel engine is known as an internal combustion engine since the flame (heat energy) is located inside the engine inside the cylinder. The external combustion engine can be seen to be an engine from the past since the efficiency of it is much less than that of an internal combustion engine. The heat energy in an ICE is better maintained inside the cylinder assembly whereas the heat energy in an ECE is lost constantly to the surroundings.

The fuel for the Mamod steam engine is placed underneath the boiler on a fuel tray. The typical fuel used for Mamod engines range from solid fuel tablets, methylated spirit, gel to the more modern gas burner.

The heat from the fuel increases the temperature of the brass boiler which is a good heat conductor (transfers heat energy with little loss) and therefore increases the temperature of the water inside the boiler. For those interested in the world of physics and thermodynamics:

$$pV = nKT$$

Where, for our example of the water inside a boiler, p = pressure of water*, V = volume of water inside the boiler, n = number of water molecules, K = the Boltzmann constant* and T = temperature of the water. The volume of the boiler, number of water molecules and Boltzmann constant are all constants. Therefore, this means the pressure of the steam is proportional to the temperature. If the temperature of the steam increases, the pressure it exerts on the boiler will also increase.

That's the science-bit out of the way ... phew!

* Treating the water in the boiler as an ideal gas.
**Boltzmann constant = $1.3806488 \times 10^{-23}$ m^2 kg s^{-2} K^{-1}

The way a Mamod steam engine works can be broken down into steps:

1. The hot pressurised steam from the boiler is passed down copper piping connected to the cylinder/piston assembly.
2. The pressure of the hot steam causes the piston to be pushed out of the cylinder. When this happens, the cylinder's angle to the engine bracket changes, blocking off the piping from the boiler and introducing a hole which dissipates the steam inside the cylinder to the surroundings (also known as the exhaust). This makes the pressure inside the cylinder ambient (the same as the surrounding atmosphere).
3. When the piston is as far out of the cylinder as possible, there is no force that pushes the cylinder back into the piston to start the cycle again (i.e. it is single acting). The hole to the boiler is blocked off due to the cylinder rotating. What pushes the piston back into the cylinder is the momentum of the flywheel (momentum = mass x velocity). Therefore, since the flywheel has a mass and a rotational velocity, it has momentum. The flywheel keeps rotating pushing the piston back into the cylinder and also opens the exhaust hole releasing the exhaust gases from the cylinder.
4. When the piston is pushed back into the cylinder, the angle of the cylinder to the engine bracket changes which opens the hole from the boiler piping closing the exhaust hole. When this happens, brand new hot steam from the boiler is passed into the cylinder, increasing the pressure inside the cylinder pushing the piston out again.

> **Thermodynamics...Yikes!**
>
> $pV=nKT$ is a fundamental equation that is used in thermodynamics to calculate the volume, pressure and temperature of a closed system for any situation where the gas is considered 'ideal'. Steam engines work **because** of this equation since as the temperature of the water/steam increases, the pressure increases which is used to power the piston/cylinder assembly.
>
> If we really wanted to get fancy, we could use the open system equation which is:
> $\dot{Q}_{1-2} - \dot{W}_{1-2} = m [h_2-h_1 + \frac{1}{2}(c_2^2 - c_1^2) + g(z_2 - z_1)]$ to work out the efficiency of the engine. But, if you're thinking what I am, that could probably wait another day...

This all happens many times a second in a steam engine.
The way a Mamod steam engine works is universal for all of Mamod's steam engines except for the newest models such as the SP6 and SP8 which use a double action piston cylinder and a double acting slide valve cylinder (see page 55).

Some Mamod steam engines are 'superheated'. This, in essence, makes the steam extra hot as it is getting heated by the fuel twice before it goes to the cylinder. If the steam is extra hot, it is at a higher pressure providing a larger force on the piston. As well as this, it stops any condensation from occurring in the piping which would hinder the performance of the engine. For the SP range, they are not superheated. For the SE3 and early SE1/2s, they are superheated.

It is important to remember that when a steam engine is steamed up, it is working under pressure and with pressure and heat comes danger. At any time, there must be sufficient water in the boiler so that there is enough steam produced from the heated water to keep the engine running. If there is not enough water in the boiler, it can result in the steam engine melting.

If the heat energy is not going to the water because there is not much water left in the boiler, the boiler will start to heat up to very high temperatures, creating a new risk of the solder joints and boiler melting damaging the engine (as well as the risk of the boiler exploding). A boiler full of water should only be steamed up once on one burner tray full of fuel (with each steam up lasting around 5-7 minutes).

After every steam up, you will need to refill the water within the boiler although, to be sure, you should regularly check the water level in the boiler since some engines are more water-hungry than others (such as the engines with more pistons). This can only be done on engines with the sight glass fitted.

Steam engines, in general, are not efficient in converting the energy from fuel to kinetic energy. Energy is lost mostly to heat from the boiler and hot steam exhaust gases. As well as this, all steam engines have gas leaks in them which will reduce the general pressure build up. This mainly occurs around the piston area.

Below is a look at how the TE1a steam tractor works (the way Mamod engines with this piston/cylinder configuration work are all exactly the same).

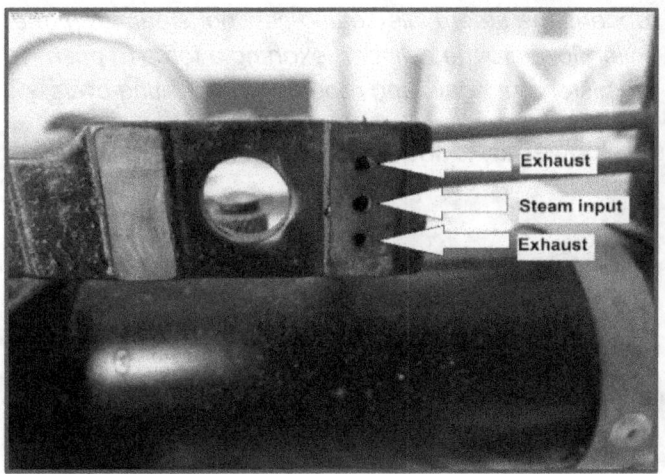

The engine bracket here has three holes in it. The middle hole is the steam input from the boiler with the other holes being exhaust pipes to the chimney.

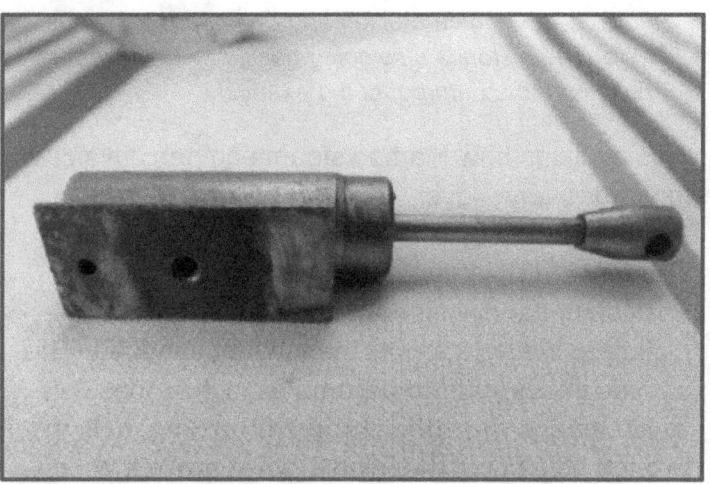

The cylinder has two holes. The more central hole is a thread which holds the cylinder to the engine bracket. The smaller hole to the left is for the steam input and exhaust.

The hot steam in the boiler travels in the pipe and goes through the middle hole in the engine bracket and into the hole to the left of the cylinder. This steam forces the piston out of the cylinder.

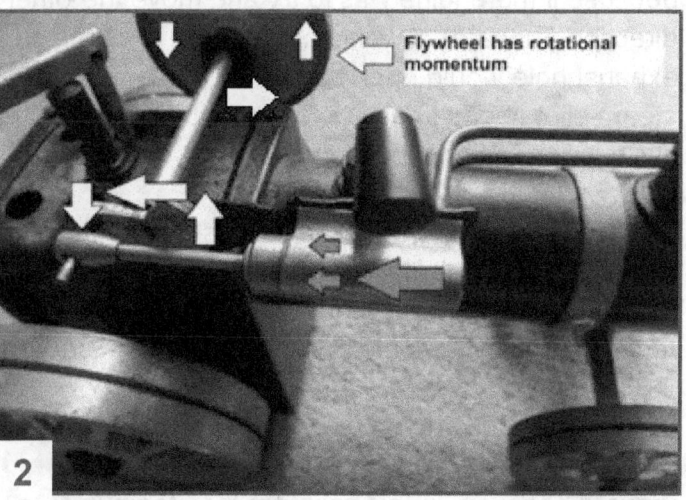

The piston has been forced out as far as possible causing it to rotate around the crankshaft making the flywheel rotate. This means the flywheel has rotational momentum.

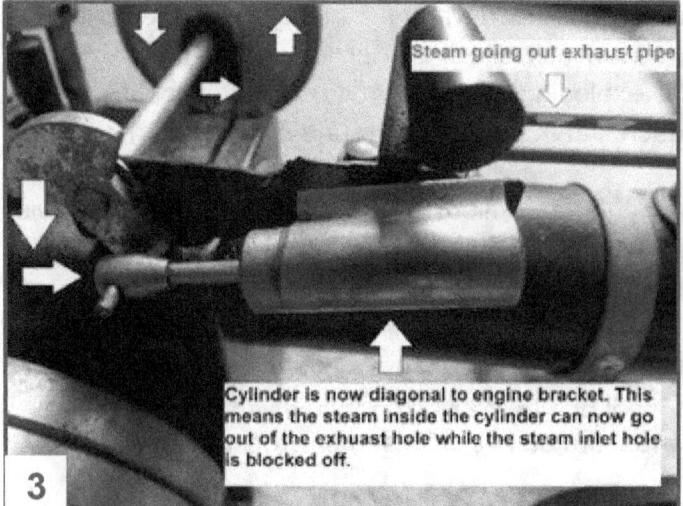
3
Steam going out exhaust pipe

Cylinder is now diagonal to engine bracket. This means the steam inside the cylinder can now go out of the exhaust hole while the steam inlet hole is blocked off.

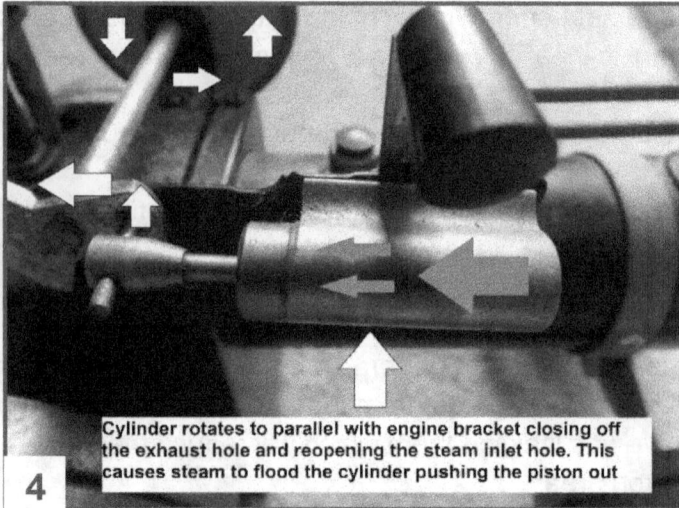
4

Cylinder rotates to parallel with engine bracket closing off the exhaust hole and reopening the steam inlet hole. This causes steam to flood the cylinder pushing the piston out

The rotational momentum of the flywheel makes the crankshaft continue rotating. This causes the cylinder to rotate which blocks off the steam inlet hole on the engine bracket and open up the exhaust hole. The piston is also going back into the cylinder pushing the steam out of the exhaust.

The piston is pushed back into the cylinder causing the cylinder to become parallel with the engine bracket again. This closes the exhaust outlet and reopens the steam inlet letting new hot steam from the boiler flow into the cylinder, exerting a force to push the piston out again and repeat the oscillating cycle.

This explains how Mamod steam engines physically work. However, it does not explain how the forward/reverse lever works.

The forward/reverse lever works by adjusting the position of the cylinder on the engine bracket. The picture to the right shows the lever pushed left. This causes the cylinder to become as high as possible which means when the cylinder is rotating, only the two top holes on the engine bracket are in function. This results the engine rotating in a way which would make it reverse.

It can only reverse with the lever in this position because if the engine was to try and move the other direction (forward), it would not be able to as the exhaust hole would open before the steam inlet. The steam inlet needs to always open first in an oscillating cycle for the engine to run.

The next picture shows the lever pushed as far to the right as possible. It is clear that this has caused the cylinder to move lower down the engine bracket resulting in only the bottom two holes being used.

For this reason, when the engine tries to run, it can only run in a direction to make the engine move forward because, in this case, for the steam inlet to flood the cylinder with steam before exhausting out of the exhaust hole, the crankshaft must rotate in a forward motion.

The Double Acting Slide Valve Cylinder with Slip Eccentric

For newer engines that date from around mid-2000s onwards, the traditional basic piston/cylinder assembly was replaced with a modern assembly also known as the double acting slide valve cylinder with slip eccentric. It is a handful, I know. However, hopefully the following pages will enlighten you to understanding this new cylinder and how it operates.

A double acting slide valve cylinder is basically a cylinder which allows steam to enter on both sides of the piston. On the traditional piston/cylinder assembly, steam can only enter the cylinder from one direction, applying an outwards force to the piston. On a double acting slide valve cylinder, steam can enter in front of *and* behind the piston, providing two forces on the piston: one pushing it out and one pushing it in hence the term 'double acting'.

This type of drivetrain can best be described through the art of diagrams:

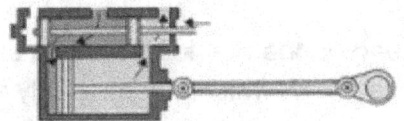

Fresh steam from the boiler enters through the middle of the slide valve and makes its way into the cylinder.

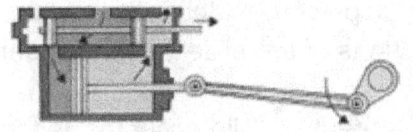

The steam forces the piston to the right, squeezing the used steam from the previous cycle out of the cylinder and slide valve housing and out of the system.

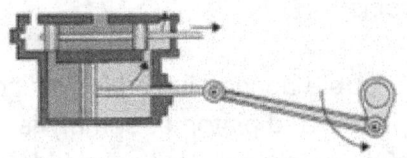

The slide valve starts to move to the right, closing off the inlet of steam from the boiler as well as closing off the exhuast.

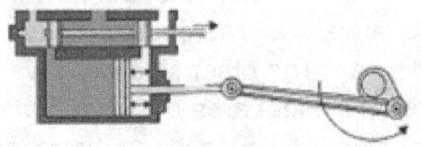

The steam inlet and exhaust are now closed off by the slide valve. The flywheel's momentum keeps the cycle going.

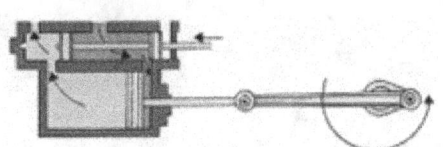

The slide valve moves to the left due to the rotation of the flywheel, letting fresh steam from the boiler flow to the right of the piston head now, forcing it to move to the left. The steam used on the previous cycle can now escape to the top left. The cycle repeats.

From the diagrams on the previous page, it is should be much clearer to how a double acting slide valve cylinder works. It relies on **one** pipe from the boiler to the cylinder (which is the middle central hole at the top of the cylinder). The two holes to the upper left and right of the cylinder are the exhaust holes – there are two since there are two power strokes to every cycle. A double acting slide valve cylinder can only work when there are two rods connected to the flywheel. One rod must connect to the piston head while the other rod connects to the slide valve.

In essence, the double acting slide valve cylinder is an improvement from the traditional piston/cylinder assembly because:

- The motion can be easily reversed
- It provides much more torque than the traditional piston/cylinder assembly can
- There are double the power strokes to every cycle
- It can self-start
- It is simple to operate, reliable and customizable (change the timing of the slide valve to exhaust gases sooner/later for optimum performance)

Saying this, because there are two power strokes to every cycle instead of one the double acting slide valve cylinder uses much more steam than a single piston/cylinder assembly, which makes it only a good source of power for the larger Mamod engines with larger boilers.

Since the piston head is totally hidden within the cylinder, it is a necessity to oil up a slide valve cylinder and piston. This problem of accessing inside the cylinder to oil the components up had been dealt with by Mamod through including on engines with a slide valve cylinder a displacement lubricator, which enables oil to go into the cylinder assembly to reduce the internal frictions of the slide valve cylinder.

Some of Mamod's double acting slide valve cylinders come with slip eccentric. 'Slip eccentric' is the term that usually defines a slotted cam which automatically changes the timing of the slide valve so the engine can run in reverse. The only engines that have a slide valve cylinder with slip eccentric are the mobile engines, such as the Centurion, Samson and Showman's Special.

Although there are many benefits (bullet pointed above) to using the double acting slide valve cylinder with slip eccentric, there are also some negatives. The cost of the cylinder and piston assembly is much higher than that of a basic piston and cylinder assembly. Therefore, if anything does fail on a slide valve cylinder, the cost of repair will be much higher.

As well as this, the double acting slide valve cylinder will always inherit the problem that the power strokes do not provide the same force on the piston head. This is because one side of the cylinder will have nothing but pressurised steam in it to push the piston head one way. The other side will have pressurised steam **plus** the piston rod. Since the piston rod will take up a slight area of space that could have been filled with pressurised steam, the overall pressure on that side will be less than the other side making the power strokes uneven. For a Mamod engine, this is not a problem at all. However, for large full size steam engines that use the slide valve cylinder, this can become quite a problem when it comes to gaining maximum torque from the engine.

General Look - Mamod Stationary Engine

The stationary engines can be seen to be the heart of the Mamod brand since Geoffrey Malins first created these engines back in 1937. Ever since then, Mamod have continued the stationary range from the SE range to the newer SP range. Either way, the stationary engines by Mamod always follow the same structural design:

- A base. For the older versions, this will be flat. For modern versions, they are slightly raised. The base is always painted red for all stationary engines (except for the special edition Forest Classic SP6 with a green base).

- A boiler mounted onto a firebox of some kind. For old engines (SE range), this is a black firebox. For modern engines (SP range), the engine is mounted around chrome which is held together by a black surrounding.

- An engine bracket mounted to the base of the engine which has the piston and cylinder assembly on it. For larger engines such as the SE3, two engine brackets are on the base to accommodate two pistons. The boiler is connected to the cylinder via copper piping. The piston is connected to a crankshaft which houses the flywheel. This enables the engine to power other Mamod accessories, as the flywheel can have a drive band connected to an adjacent pulley, which powers the Mamod accessories.

If you look at any Mamod stationary engine, it will always have this same structure. This makes it easier to restore stationary engines because many of the parts are exactly the same for different engines.

A stationary engine is also a nice engine to first start out with when it comes to restoring your first Mamod. They are generally not as complicated as the traction engines due to the fewer parts. As well as this, the colour combinations are not as complicated as the traction engines. There are three colours for the old stationary engines (SE range) being red, green and black and four colours for the new stationary engines (SP range) being red, blue, black and grey (the grey only applies to the SP4 - the raised grey platform).

The painted parts to an SE engine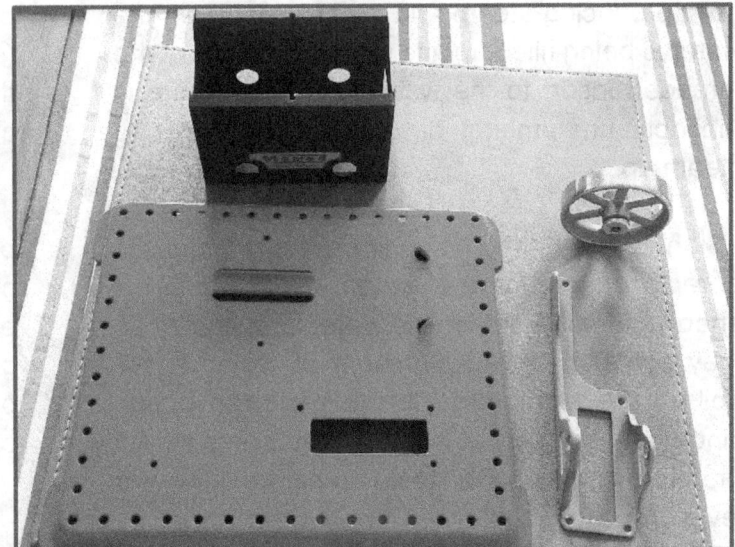

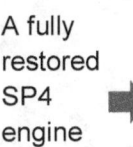 A fully restored SP4 engine

Below are the parts of the stationary engine and their jobs:

- Boiler. The boiler is a cylindrical brass container that holds the water. Being gas tight, when the water is heated, the pressure inside the boiler increases which is used to push the piston out of the cylinder.

- Base. Keeps all the components of the stationary engine in a fixed position limiting any vibrations that may occur while running.

- Safety valve. This should be able to move freely up so that when there is excess steam is in the boiler, the valve will move up to reduce the pressure. This stops the boiler from potentially exploding.

- Water level plug. The water level plug does not appear on all stationary engines as some engines have a sight glass instead. The point of the water level plug is to stop the user overfilling the boiler with water (overfilling will stop the boiler from producing steam and pressure). When the water is being filled up and starts to run out of the threaded insert to the water level plug, there is the optimum amount of water in the boiler to steam up.

- Sight glass. For some engines, a slight glass is a replacement for the water level plug. This enables the user to see the water in the boiler so they can put the right amount of water in the boiler. The sight glass is the preferred option since you can see how much water is left (and therefore can stop the steam up before the water level drops too low).

- Crankshaft. An important part to the engine, the crankshaft converts the linear motion of the piston into a rotary motion.

- Flywheel. Connected to the crankshaft, the flywheel gives the engine rotational momentum - it stores energy during the engine cycle.

- Whistle. This appears on the top of the boiler for the SE2/3 and SP3/4 engines next to the safety valve. The whistle uses the pressure of the steam to create a high pitch sound when the lever to the whistle is pushed down. The whistle is a common element to most Mamod steam engines.

- Regulator. On some stationary engines, there is a regulator to control the amount of steam that goes to the cylinder (and therefore controlling the speed of the engine). On the SE3 and SP6/7/8 models, the regulator is on the boiler while older SE2s have it on the cylinder block.

- Forward/Reverse Lever. Some stationary engines have a forward/reverse lever on the cylinder/engine bracket to control the height of the cylinder which influences what direction the cylinder/piston rotates. This changes the direction the engine is running and its speed (since the lever can move the cylinder so that the steam inlet hole is only partially open to the cylinder).

- Piston and Cylinder. The steam is fed from the boiler through copper piping to the cylinder. The cylinder has specific holes in it so it which enables the steam from the boiler to flood the cylinder forcing the piston out from the pressure of the steam, rotate around, and then be forced out again (and so on).

- Pulley. The pulley makes it possible to connect other accessories to the engine such as the workshop accessories.

- Engine bracket. Houses the piston/cylinder assembly and the crankshaft/flywheel/pulley assembly.

From the parts listed above, I hope it is quite clear that the stationary engines produced by Mamod are very basic: only the parts needed are used. This is why many people choose to restore stationary engines because of their simplicity.

Crankshaft
Chimney
Flywheel
Engine bracket
Safety valve
Cylinder
Piping (just one pipe)
Chrome boiler band
Boiler
Firebox
Base

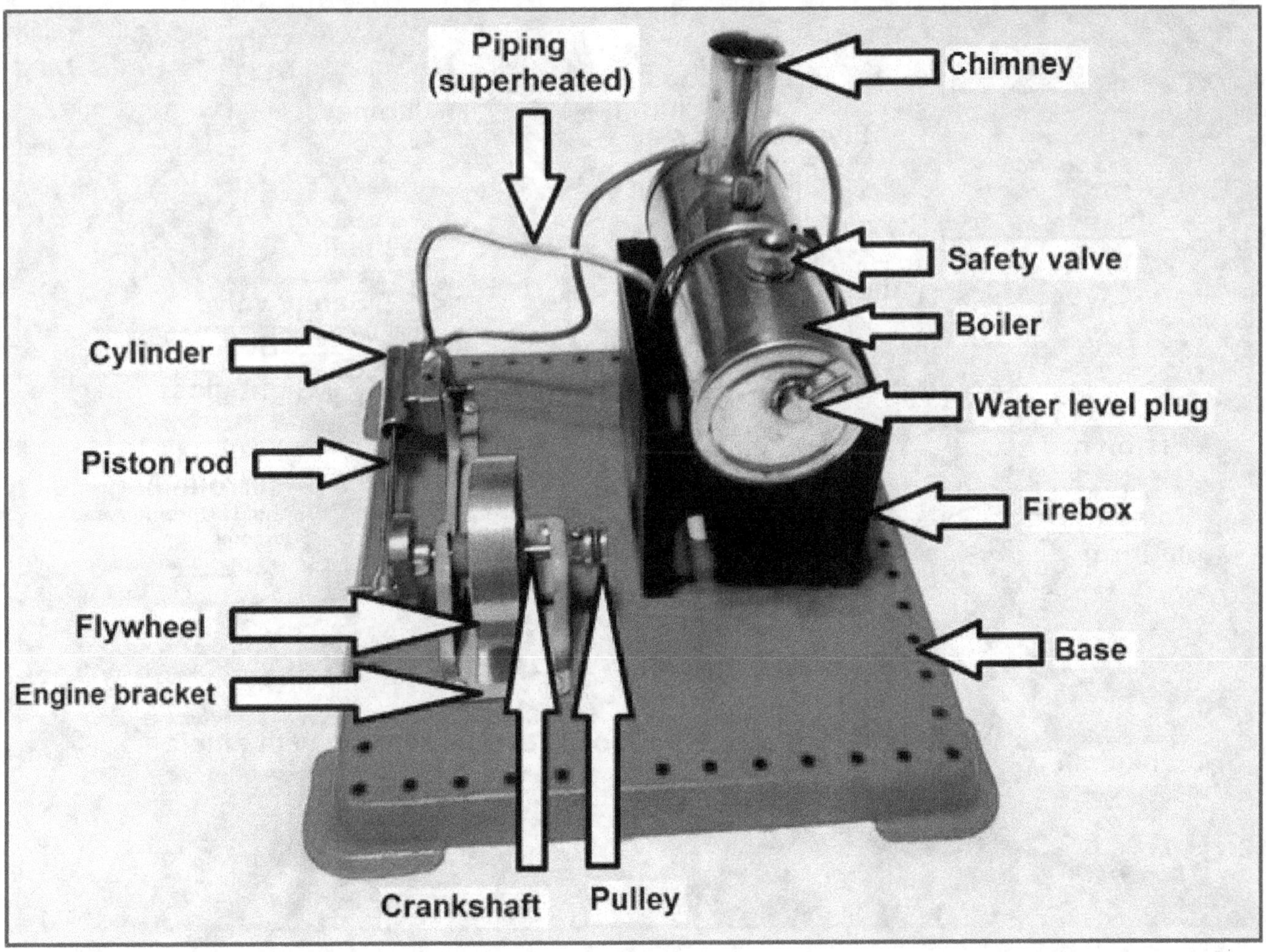
Piping
(superheated)
Chimney
Safety valve
Boiler
Cylinder
Water level plug
Piston rod
Firebox
Flywheel
Base
Engine bracket
Crankshaft
Pulley

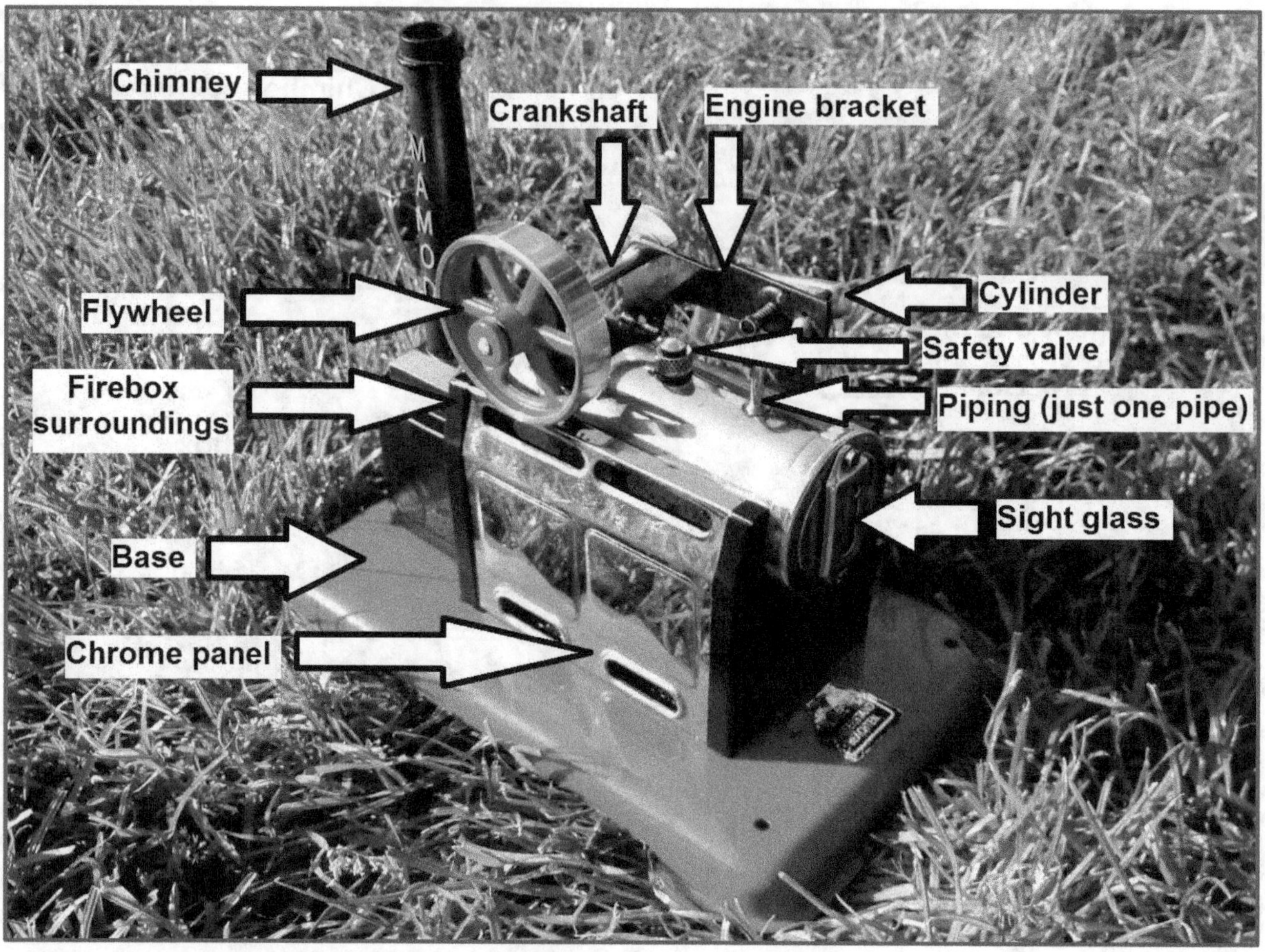

Chimney
MAMOD
Crankshaft
Engine bracket
Flywheel
Cylinder
Safety valve
Firebox surroundings
Piping (just one pipe)
Sight glass
Base
Chrome panel

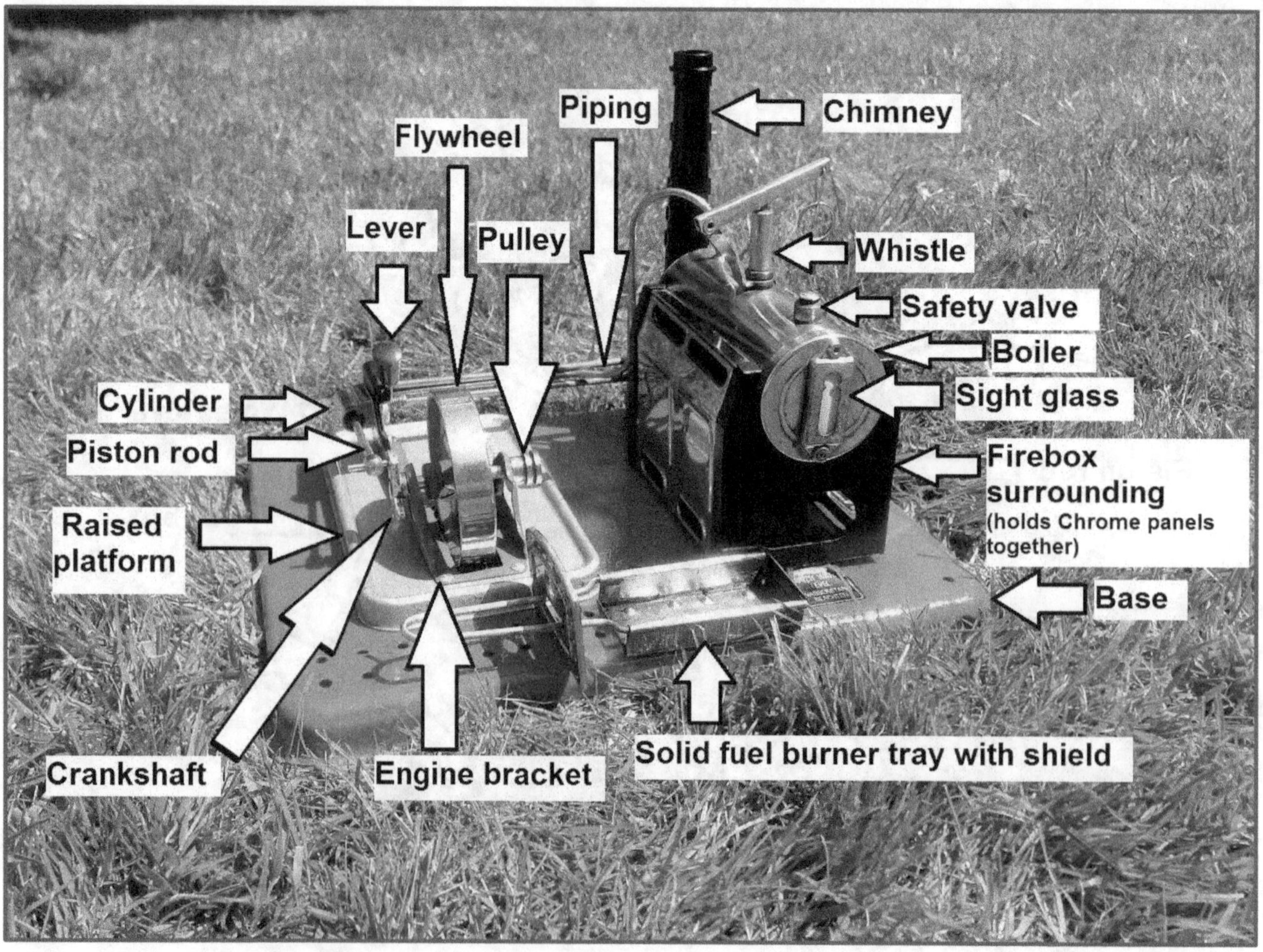

Piping
Chimney
Flywheel
Lever
Pulley
Whistle
Safety valve
Boiler
Cylinder
Sight glass
Piston rod
Firebox surrounding (holds Chrome panels together)
Raised platform
Base
Crankshaft
Engine bracket
Solid fuel burner tray with shield

General Look - Mamod Mobile Engines

The mobile engines, by Mamod, were the most popular engines by Mamod as they were their first engines that were able to move under steam power. As we have already looked at in 'The Range Of Models' P20, there is a wide variety of mobile engines which are considered to be much more complex than the stationary engines. This is because the Mamod mobile engines have a chassis, wheels, hub caps, roof, bodywork and more which enable the engine to move. To a restorer, this means there are going to be more parts to restore with a wider variety of colours and restoration techniques to use.

The way the Mamod mobile engines work is exactly the same way that the stationary engines work. For that matter, the way all of the Mamod steam engines work follow the same thermodynamic principles. The main obvious difference is that the kinetic energy produced by the rotating flywheel is converted into useful energy through a drive band connecting the flywheel to one of the wheels on the road. Due to friction between the drive band and the wheel, the wheel rotates making the whole engine move.

Most mobile engines, to get going, need to get the flywheel rotating with the road wheel *not* touching the ground. The reason for this is that to stop the engine from stalling, the engine needs to provide lots of torque. However, the torque is never higher enough to get the mobile engine to move from stationary. So, we need to get the engine revolutions up to increase the power coming out of it. This can be done by giving the engine a nudge along the ground or by lifting the drive road wheel (connected to the engine) up so that it is rotating freely in the air. When the engine is at its maximum revolutions, lower the engine back onto the ground. The engine will stagger a bit but should keep on going. I tend to find the best way to steam up a mobile engine is by disconnecting the flywheel to the drive wheel by removing the drive band like I have done below (so that the engine doesn't move).

The components that make up most of the mobile engines by Mamod are all the same when looking at the engine. The only difference is the linkage of the flywheel to the rear wheel enabling the engine to move forward as well as the obvious differences in the bodywork of the models.

My SW1 Steam Wagon steaming up without the wagon body kit on. ➡

Canopy
Engine bracket
Lever
Whistle
Chimney
Piping
Flywheel
Safety valve
Driveband
Hubcap
Front fork
Boiler
Leaf spring
Firebox
Chromebox

Engine bracket
Flywheel
Chimney
Crankshaft
Chromebox
Scuttle
Piping
Driveband
Screw for steering rod
Front Fork
Hubcap
Front Rollers
Safety valve
Firebox

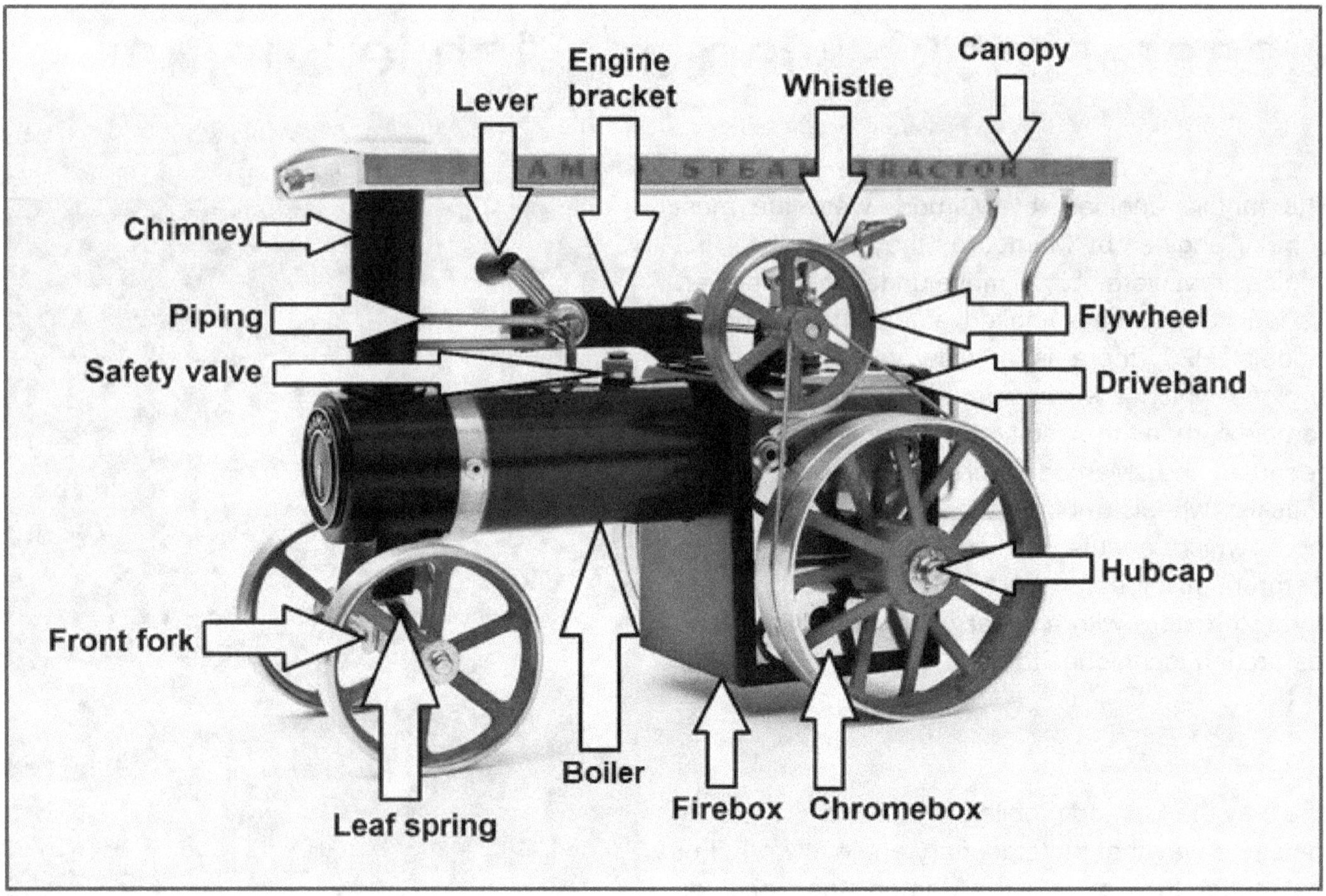

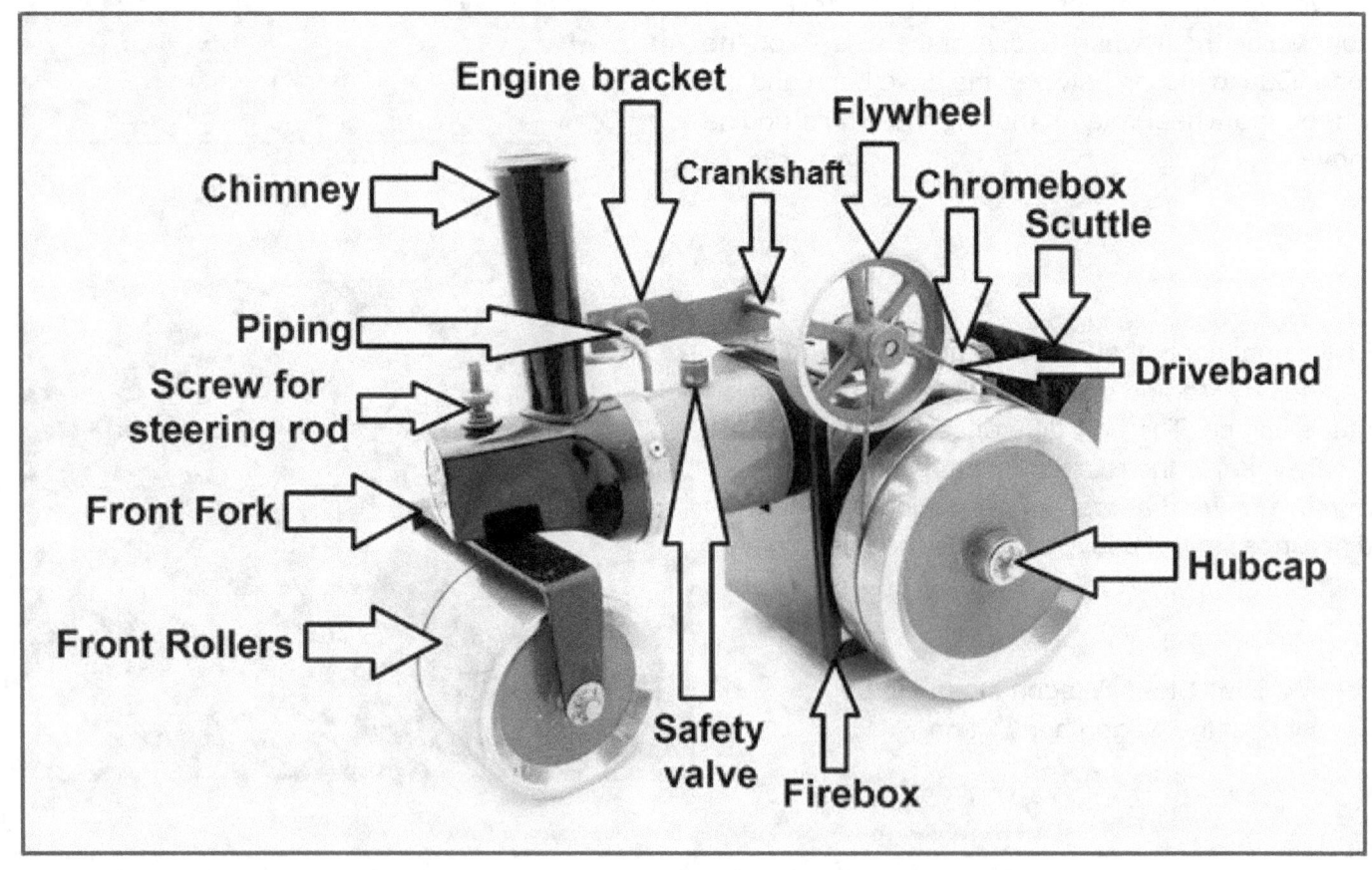

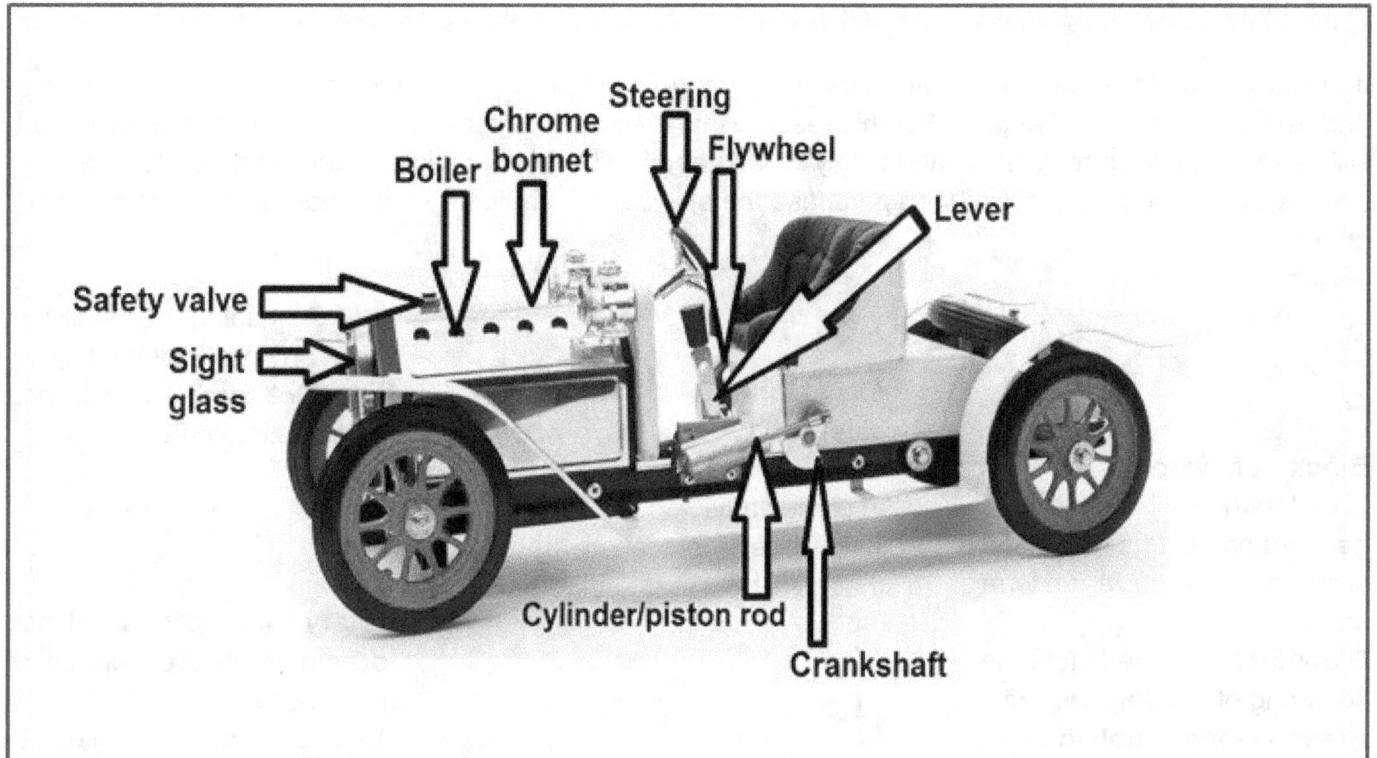
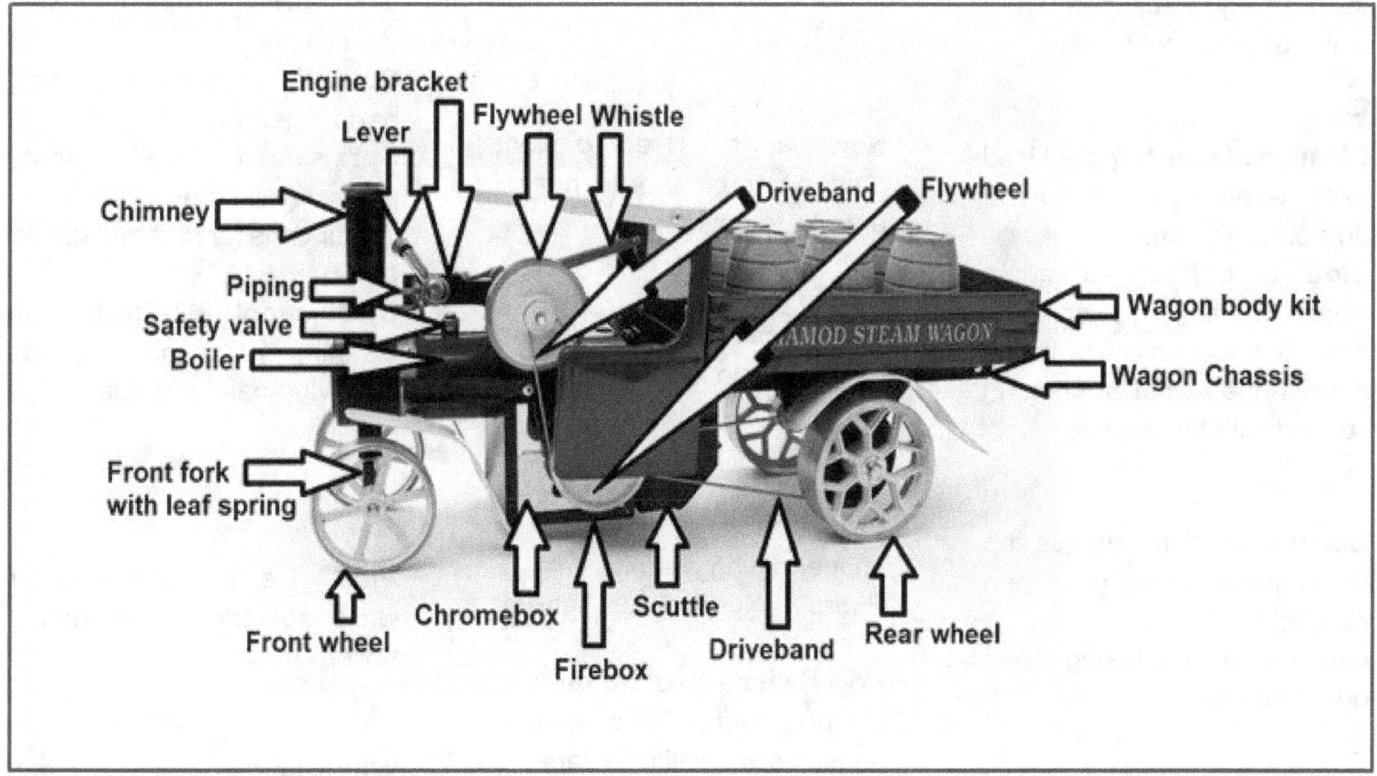

Equipment/Tools Needed For a Restoration

If this is your first restoration, you are most likely not going to have the equipment needed to take apart and restore your Mamod engine. For this reason, this chapter will go into detail with what equipment you will need during your restoration and why you will need it. Therefore, by identifying what needs to be done with your engine, you can then purchase the appropriate equipment/tools needed to restore your engine.

A

B

Block of wood - Always useful when it comes to hammering things out and sanding down in a uniform manner.
Blowtorch - Used for the soldering of the boiler/piping.
Brasso metal polish - A brass polish used for polishing metal parts.
Bucket of water/sand - Used as a safety precaution when doing anything with fire.

C

Clear sealable bags - Used to store loose parts in so you do not lose them.
Cloth or old rag - Used to clean parts by applying Brasso metal polish to it and protect metal parts when in contact with vice/pliers.

D

Decal - Used to replace the decal on parts that have been painted.
Drill - Used for taking rivets out of holes.

E

Emery paper - Used to clean different areas of the engine (get a range of emery paper from 200-1000 grading).

F

Fire blanket - Used as a safety precaution when dealing with methylated spirits in particularly.
Flux - Used for cleaning areas that are about to be soldered.

G

H

Hammer - Used to hammer parts out such as end caps on boiler.

I J K L M

N

Newspaper - Used to paint parts on so paint does not go anywhere unwanted.

O

Oven (optional) - Used as a quick way to dry just-painted parts.

P

Pipe cutters – Used to cut copper piping cleanly.
Pliers - Needed for general use.
Pop rivets - Used to replace old pop rivets. On a Mamod the sizes you will need are:
- 3/32" Aluminium rivets for SE engine brackets to base and Firebox to chromebox on mobile engines.
- 3mm Copper closed end rivets for the sight glass and boiler to firebox.
- 1/8" Aluminium rivets for just about any other rivet fitting.
Pop rivet gun - Used to pop the rivets into holes.

Q

R

Rotary tool (optional) - Used to help clean the surface of metal parts.
Rust Converter - Used to treat rust on rusted parts.

S

Screwdriver - Used for taking out screws.
Solder - Used for connecting parts together.
Spray paint - Used to spray parts (will need different colour cans for different colour parts).
Steel wool - Effective in cleaning solder joints up and removing paint and rust.

T U

V

Vice - Used to clamp boiler while soldering and heating boiler up.

W

White spirit - Used to clean parts before painting.
Work surface - Used as the surface to restore the engine on.

X Y Z

Restoration No Nos!

Before we get into the restoration process for your engine, it is important to do these restorations correctly since applying the wrong techniques to the restoration will result in a poorly restored engine that would have looked better if you left it alone!

Saying this, there is not a right or wrong way to restoring an engine. Every restoration will provide different challenges the restorer to overcome. The reason for this chapter is to try and prevent you from making restoration mistakes so that your engine looks as good as it could possibly be.

Paint Your Engine the Right Way

A huge element to how good your engine looks will be based on how you have painted it. With painting your engine, there are a few rules to always abide by;

- Always spray and never brush since brushes will make lines in the coat.
- Don't overspray since this can lead to drip marks on the surface.

- Always clean the surface being painted with white spirit and give surface time to dry.
- Spray with the right type of paint for the part (e.g. high temperature).

Don't Hide a Problem with Your Engine

If you find a problem with your engine, there is a tendency to try and cover up the problem or temporarily solve the problem. However, this is bad! Whenever you find a problem with your engine, always fix it so it is 100% A-Okay. You are going to find problems with the engine you are restoring - fact. As a restorer, you need to learn to confront these problems and tackle them correctly one at a time.

Be Consistent with Your Restoration

Another mistake that could happen when you restore your engine is that you are not consistent with your restoration:

- Use the same size screws or rivets for the whole engine.

- Put the engine back together the same way you took it apart in the same order. Take pictures as a precaution.

- Use exactly the same shade of paint for the same colour parts.

- Clean the metal parts the same way to produce the same shine.

The only time you would not use the same size rivets would be if you had a sight glass on your engine which usually requires slightly larger rivets. However, the more consistent you are with your engine's restoration, the better it will turn out to be.

Don't Think You're an Expert

There is no greater risk to damaging a restoration project than to think you are an expert at restoring your engine. When you have to do something on your engine, always think logically about what you have to do. Are you using the right materials/liquids/tools for the job?

This even applies for me. I am not an expert on Mamod engines. The only true expert there is on Mamod engines is Geoffrey Malins. He produced 567 engines in his first year alone with all of them being hand built. If ever you find one of these engines, you will be surprised that they all still work over 80 years on even if he did say they were *'the worst I ever made as I had to begin at the beginning and find out everything'*. Mamod steam engines are great pieces of kit that can help anyone learn about the fundamental principles to engineering and steam engines. Only once you have started and maintained a successful live steam toy business from scratch can you truly call yourself an expert! Since that is unlikely to happen for all of us, we still have the capability to learn from these amazing pieces of engineering from the past and present.

Universal Restoration Tips

Since many of the Mamod steam engines undergo the same processes when being manufactured and the designs of them are quite similar, there are some universal tips that can apply to every type of engine Mamod makes. Therefore, this chapter will be extremely useful to you during your restoration.

Painting Your Mamod Parts

Every Mamod engine has some part/s that are painted. Therefore, when it comes to restoring these parts, it is important to know how to repaint them. If done correctly, the restoration of a painted part can have the potential to look just as good as new.

Prepare the Surface

The first part to repainting a part is to prepare the surface. You need to make sure the part is as smooth as possible as any imperfections, lumps or chips will become amplified when painted. There are a few options you can choose to making the part's surface as smooth as possible.

- Strip the whole part of paint completely. If the surface is that bad, you may find it easier to just remove all the paint. This is most common with parts that have rust on them. To completely treat the rusty parts, you will need to remove all of the paint and visible rust (with white spirit and emery paper) and then apply rust converter onto the surface. Be careful removing the old paint since some of the old engine paint may have contained lead - make sure you are in a well-ventilated area. It is advisable to wear a suitable mask so you do not breathe in any of the dust.

- If there are small areas of imperfections, you may want to choose to smooth that area off only. You can do through gently rubbing the affected area with fine emery paper. In this case, you do not have to remove all of the paint down to the metal. All you need to do is make the surface smooth (which you will be able to tell by running your finger over the affected area).

Rust, rust and more rust

It is extremely common for the painted Mamod parts to have rust underneath the layers of paint (remember that some of the engines are over 80 years old)! If there are any blemishes (in particularly, raised lines) in the paintwork, it is a good guess that there is rust under the paint's surface and it will keep spreading. Therefore, you will need to re-strip the paint to remove the rust to prevent it from spreading.

However, the 'Rust, rust and more rust' box on the previous page states that some of the imperfections are caused by a build-up of rust underneath the surface of the part. When deciding whether to smooth off the surface or take all of the paint off, you need to identify if the imperfection is caused by rust underneath the surface of the part or if it is just a simple imperfection from general wear and tear.

For the best restoration, it is essential to completely derust all of the painted parts. The chances are that over the years, rust has got underneath the part's paint and rust, on metal, spreads like a virus. You can remove all of the visible rust off with emery paper. However, there will be rust that you cannot see embedded into the metal that will need neutralising by applying rust converter to it. You can then smooth off the surface and spray over it. Remember that any imperfections on the surface before spraying will still be noticeable after spraying.

Spray Paint the Surface

You are now at the stage where you are ready to paint the part since the surface is completely smooth. Before you do so, there are a few things you need to do:

- Some areas of a part do not require them to be completely painted (such as the flywheel). Therefore, it is a good idea to cover the parts you don't want to paint with masking tape. For the flywheel, you can mask off around the outside of the wheel. Of course, you don't have to. But, it will save you the time and hassle removing the paint from the areas you did not want to be painted.

- Before you paint the surface, you should rub the part over with a cloth that has been lightly soaked in white spirit. This will remove any microscopic dust on the part. Be careful when using white spirit. It is a dangerous chemical that is irritant to the skin, dangerous to the environment and extremely flammable. After use, always wash your hands with soap.

- Make sure you are painting in an area that is well ventilated. It is a good idea to place the part on newspaper so that no mess is left after spraying. As well as this, make sure the area you are spraying in does not have any dust in the air as this will settle onto the metal producing imperfections on the coat of paint.

Once you have done all this, you are ready to paint. Make sure you spray with a fluid and consistent motion *across* the part. If you spray directly onto the part, you risk forming drips on the part from over-spraying on a specific area. *It is better to do lots of light coats than a few heavy coats.*

What we also need to bear in mind when it comes to spray painting a part is to choose *what type of paint* to use. What parts of the engine get hot and will need heat resistant paint? Using normal paint on an area where it will get hot will cause the paint to bubble and become soft affecting the aesthetics of the engine. If you want the best surface finish and paint adhesion, it is best to use a good primer before spraying your top coat since primer will smooth out imperfections and bond better with the metal.

(Images left to right: SE2 flat base having primer applied before being sprayed red; same SE2 flat base being sprayed red after primer coat; SE2 flywheel being sprayed on newspaper with masking tape around the outside of the wheel; sprayed firebox and engine bracket drying in the sun)

 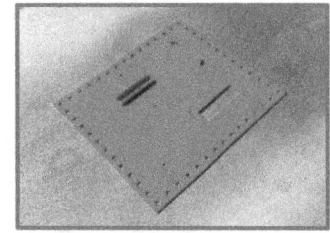  

Cleaning the Metal

With Mamod steam engines, there are four main types of metal you will be cleaning (and how to identify each metal):

- Brass - A shiny golden colour that tarnishes to a dark gold over time.
- Steel - A dull grey colour which could have dark specs of rust on it over time.
- Chrome - Chrome is shiny and mirror-like making it easily identifiable.
- Copper – A light auburn (brown/red) colour which darkens when tarnishes.

> **Important Tip!**
>
> The cylinder and piston are always made from brass on Mamod engines. This is because brass is a self-lubricating material (therefore, it is perfect for using when a piston is constantly moving in and out of a cylinder). It is *extremely* important that you do not clean the piston head or the inside of the cylinder at all as these will have 'lapped' together to produce a good gas tight seal.

There is also a few other materials Mamod have used over the years such as Mazak. However, these materials can be treated the same as steel as they were replacements to steel.

Cleaning Brass

Brass is a great material to clean because it does not rust. When brass oxidises, the surface of the first layer of brass oxidises producing a tarnish effect. This layer of tarnish preserves all of the other layers of brass underneath the surface. Thus, when it comes to cleaning brass, all we need to do is remove this oxidised layer from the brass to recover the full shine. This can be done in a few ways.

Emery Paper

For some engines that have lots of tarnishing and lime scale on the boiler, you can use emery paper to clean the brass. The benefits of this is that it will take less time than the other methods to clean the brass. However, the emery paper *will* scratch the surface of the brass. Although the scratches are not that noticeable if you use fine grade emery paper, the small scratches over the brass will make it less 'mirror-like' when finished. It is advisable to polish the brass surface thoroughly to reduce the appearance of the scratches. For this reason, if you choose to use emery paper, make sure the grain of the paper is graded 1000 or higher and that the emery paper is soaked in water to make the emery paper even less coarse.

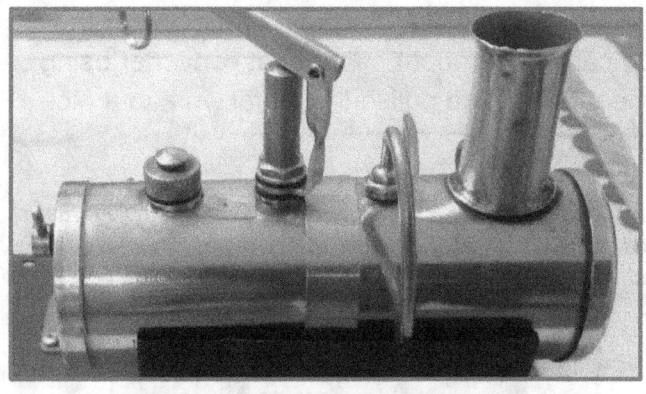

If you choose to use emery paper on your brass parts, it is a must to use brass polish and a cloth after to minimise the scratches caused by the emery paper.

Brass Polish and Cloth

This is my preferred option as it will not scratch the metal work as much as the emery paper and works pretty fast too. Brass polish has micro-beads in it which means it still scratches the surface but the scratches are at the microscopic level. Ultimately, the more time spent on cleaning brass, the better it will look.

From the pictures on this page and the previous page, I hope I have made clear the difference in shine you get from using emery paper and then brass polish and cloth to just using brass polish and cloth. Whenever possible, try to stick to brass polish and cloth when cleaning brass as it will always make it look that bit shinier. However, we don't always live in a dream world and, even with brass polish, marks on the brass sometimes will not go away. This is when to use fine emery paper and then brass polish.

Cleaning Steel

Fortunately, for us restorers, steel is a bit easier to clean than brass because when steel oxidises, it turns straight to rust. Steel is also not as shiny as brass. Therefore, you can afford to use emery paper to clean it and then use metal polish to make it shinier.

Since steel oxidises to rust, it is essential to treat all existing rust so no further rust will form. Treat rust by brushing on rust converter liquid onto the metal. The rust converter then neutralises the rust creating a black chemical barrier. After this, you can then smooth off the rust converter using emery paper and polish the metal work to a nice finish.

(The firebox on this SE1 engine might seem a complete rust bucket. However, all it takes is emery paper to remove the rust and rust converter to treat any more existing rust on it).

Cleaning Chrome

Chrome work on Mamod engines is difficult to clean since it is only a thin chrome plate on steel. This brings around the problem that if you clean it too harshly, you will penetrate the thin chrome plate to steel which, as we know already, will contrast significantly to the chrome that has remained around it. This is why if you cannot use emery paper at all on chrome.

The reasons for cleaning the chrome work on Mamod engines are because of the following:

- Rust has formed on the chrome from the layer of chrome being removed letting the steel oxidise.
- Limescale has built up from being in contact with water over the years.
- It is generally dirty.

For each of these reasons, you can use metal polish to clean the chrome work. However, be warned that the brass polish will still be slightly abrasive, as it has micro-beads in it. Therefore, do not rub too hard. If you haven't seen any improvement to the shininess of the chrome, leave it. Rubbing any longer than is necessary will remove the chrome layer showing the dull coloured steel.

Cleaning Copper

Every engine made by Mamod will use copper piping to connect the boiler to the cylinder. The reason for this is because, just like brass, copper does not rust. Since it is also more malleable (can bend easier) than brass, it is the perfect material to use as piping with hot steam inside.

Since I have described the properties of copper similarly to that of brass, we can clean it in the same way. All we need to do is remove the top layer of tarnished copper to make it look much lighter in colour and bring back the shine. Again, like with brass, be careful not to scratch the copper. Metal/brass polish is good to polish the copper piping on engines.

Be careful when you do clean the copper to not bend it. If you bend the piping you are causing the copper piping to work harden which fatigues the atomic structure of the copper. If you bend it too much, it will eventually snap. Therefore, it is best to not bend the copper piping at all (or if you do, do it gently and not too much: there are some engines that require you to slightly bend the copper piping to remove it from the other parts of the engine such as the supercharged SE range).

If you do snap the copper piping or the pipe comes out of its soldered socket, you will need to resolder the piping back into the socket. This is okay if it has just popped out. However, if you have snapped the piping, you need to make sure that the piping has not closed in on itself preventing any gas from escaping (consequently, your engine won't work). Therefore, always make sure to file down the end of the piping affected and remove all burrs so that there is a gap for steam to continually flow through the piping.

Why does copper piping snap when bent? (WARNING - lots of material science below)

Copper, like most metals, has a crystalline structure. This means that when you look at copper to an atomic level, all of the atoms of copper are arranged lattice-like uniformly. When the copper piping is bent, the structure of the copper changes and dislocations are created (irregularities in the structure where atoms are out of their position in the crystalline structure) – this is known as work hardening. This is because dislocations are generated and move when a stress is applied to the copper piping (i.e. the copper piping is bent). Although dislocations make the metal harder since dislocations make the structure more densely packed and make it harder for movement in the atomic structure to occur, dislocations also make the material much more brittle. When dislocations all move to one area (i.e. the part being bent on the copper piping), the movement of ions (electrically charged atom due to gain/loss of electrons) is harder reducing the force keeping the atoms together (which means the copper at that point becomes more brittle). Therefore, when working with the piping in Mamod steam engines, do not bend the piping as this *will* make it more brittle.

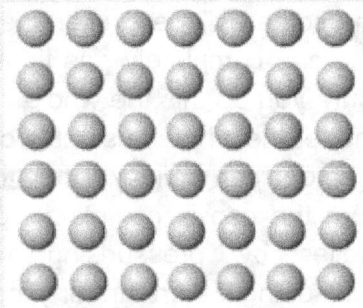

⬅ Crystalline structure without any dislocations on weaknesses

Crystalline structure with a dislocation (this type of dislocation is known as an edge dislocation or line defect). ➡

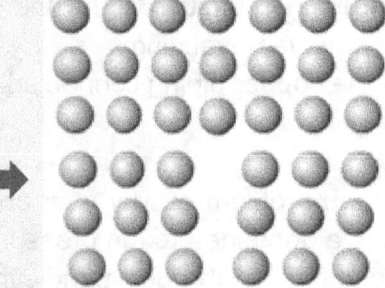

Phew, glad we got that out of the way!

There is also the process called annealing, which involves slowly heating up and cooling a metal to help get the structure back to what it originally was: crystalline. This can be applied to copper piping to help the bent pipes become ductile again since the heat allows dislocations in the structure to revert back to the lattice structure.

Restoring the Boiler

The brass boiler on your Mamod steam engine is the most important aspect to the whole engine. It is where all the pressure is produced from heating the water. As well as this, a nicely restored boiler can simply change the whole look of an engine and make it look magnificent. Therefore, it is important you know how to restore your Mamod's boiler.

In this section, I will look at the two types of restoration for a boiler: the aesthetics and the mechanics.

Mechanics

If you find that your boiler is faulty and it is not working, do not panic! Although this is a serious problem regarding your engine working, any problem can be fixed - it is just a matter of time, effort and money.

Firstly, let's understand what the boiler's job actually is on the engine. After being filled with water, and heated, a proportion of the water inside the boiler evaporates into steam creating a pressure inside the boiler. It is this pressure that exerts a force on the piston/cylinder assembly to create kinetic energy (movement). Therefore, the boiler needs to be completely gas tight so that no pressure is lost. The main reason why a boiler could be faulty is because there is a leak which is allowing steam to escape from the system. For this reason, let's list the areas from which steam can potentially be lost from the boiler together with ways of fixing these areas:

- The front or rear end cap.
- Threaded inserts such as for the safety valve, whistle or water level plug.
- From the copper piping connected to the boiler.
- The sight glass.

To find the problem of why your boiler isn't gas tight, it is a good idea to run pressurised air from an air compressor through it and either cover the engine in soapy water or dip it in a bucket of water. The bubbles appearing will help you locate where it is not gas tight.

The Front or Rear End Cap

The way Mamod has created the boiler is by pressing a brass cap onto the front and rear end of the cylindrical boiler and using solder to fill in all the air gaps. If your caps are leaking air, or are in a bad state, the best way to solve this problem is to resolder the joint.

End cap of a Mamod SE1/2 boiler ➡

For this to happen, you will need to first take the cap off the boiler. This can be done by using a blowtorch or solder iron torch (be careful using as they are extremely hot and will make the brass extremely hot) to warm up the cap and boiler (that is clamped into a vice as shown on page 74). Once the brass assembly gets hot enough, the solder that is keeping them together should start to melt (you probably won't be able to see this as the solder will be melting inside the boiler). When the solder inside the boiler starts to melt, you should be able to slowly rotate the cap off.

To reconnect the cap to the boiler, make sure that the area where solder will go (edge of the boiler cylinder and the lip on the cap) is completely clean and covered in flux (flux is a chemical cleaning agent) – use emery paper on the joint to clean. Once you have done this, you can press the cap back onto the boiler. It will be a tight fit, but once it is in place, you can start adding solder to the gap between the boiler and cap. Due to the capillary action, the solder will be sucked inside the gap making the cap completely gas tight to the boiler. Make sure you do this the whole way around the cap and let the whole assembly cool down so that the solder can solidify. After this, check to make sure the cap is completely gas tight to the boiler by pressurising with compressed air and seeing if there are any leaks by covering the areas prone to releasing air with soapy water (if there are bubbles appearing, air is still escaping and that area will need to be re-soldered).

Threaded Inserts Such as for the Safety Valve, Whistle or Water Level Plug

The majority of times a boiler will leak air is because of the threaded inserts. Over the years, the constant use of threads and contact with water makes them erode down to a point where there is a gap between the thread and the nut. At this point, the only way to rectify this problem, unfortunately, is to replace the insert. This is done using the same process as removing the front/rear cap (as previously). Heat the insert so the solder around the threaded insert melts, then remove the threaded insert. Clean the area with emery paper and use flux around the joint. Apply solder to the joint around the new threaded insert and leave to cool down.

The boiler is such an important element to the steam engine and I tend to want to try and keep the original boiler with the restoration (which is what I have done with all my restorations to date). However, some boilers could be in such a bad state that it is worth considering buying a new boiler: especially if you do not feel confident or do not have the tools to repair a boiler. Remember that anything is restorable - it is just a matter of time, effort and money. For a boiler, the time, effort and money put into restoring it may actually be quite substantial (which is why it may be worthwhile buying a new boiler instead).

From the Copper Piping Connecting to the Boiler

Every Mamod steam engine has copper piping going from the boiler to the piston/cylinder assembly. Again, this is an important aspect to the engine as it is the piping that transfers the pressurised steam to the piston/cylinder assembly so kinetic energy can be produced from the assembly.

Unlike the two other problems (front/rear caps and threads losing air) associated with the boiler, a problem with the copper piping connected to the boiler is fairly easy to repair. The first problem that may occur is that the soldered piping to the boiler is not gas tight. In this case, all you will need to do is reheat the soldered area and possibly add a bit more solder to the area so that it is gas tight.

The second problem may be a bit trickier. Inexperienced restorers may bend the copper piping while restoring it. This is fine to do but do not do it too much since you are applying work to the pipe. Work hardening to copper weakens it and if you apply too much work to it (bend it too much), it will eventually snap (as I have already spoken about previously in detail). As well as this, bending the pipe will decrease the area inside the pipe making it much harder for the steam to flow through. Therefore, as well as weakening the pipe, you are decreasing the performance of the engine too. If the copper piping to the boiler has snapped, you are going to have to replace the whole pipe with a new one.

The Sight Glass

For more modern engines, the water level plug was replaced with a sight glass. This meant that the steam engine enthusiast could see the water level inside the boiler so that he/she does not overfill the boiler with water or run it with too little water inside. It was a great addition to engines as it reduced the chances of anything dangerous happening (such as running the engine without any water in it: with the water level plug, you don't know how much water is left in the engine).

The only problem with the sight glass is that it is another area on the boiler which can potentially leak. The sight glass is sealed by pressing a rubber seal up against the boiler using either copper rivets or screws. Once that rubber seal goes, there is nothing keeping the sight glass gas tight. The good thing about repairing this problem is that it is quite straightforward. You will have to drill the copper rivets out, replace the rubber seal and place two new copper rivets back in using a pop rivet gun.

The parts which construct a sight glass →

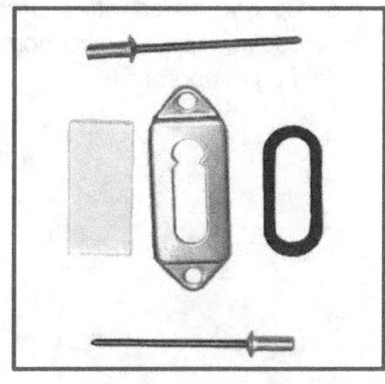

Removing the Front or Rear End Cap from the Boiler

Step 1: Clamp the boiler into a clamp with cloth around the boiler to prevent the boiler from scratching against the vice.

Step 2: Use a torch to melt the solder. Slowly pull and slide the end cap off, gently, using pliers.

Step 4: Prepare both the end cap and boiler for soldering. This is done by cleaning the inside of the cap and boiler with coarse emery paper and white spirit. Apply flux to both areas when white spirit has evaporated.

Step 4: Clamp the boiler back into the vice and slide the end cap back onto the boiler. Heat the end cap and boiler with a torch and apply solder to the whole of the end cap wall. Let the boiler cool down and remove any excess solder.

Aesthetics

If all has gone well up to now, your boiler should be mechanically stable, but dirty and tarnished. We need to now work on cleaning up the boiler to make it look as good as it did when it was first made.

We have already spoken a bit about how to clean brass at the start of the 'Universal Restoration Tips' chapter (see page 67. You can either use emery paper or brass polish with a cloth to clean it. However, with the boiler, it is the centrepiece of the whole engine. Therefore, extra effort and time needs to go into cleaning it. Here are a few rules you should try to stick to when cleaning the boiler:

- Try not to use emery paper at all. Remember that emery paper is scratching the surface of the brass. Try to stick to brass polish and cloth as this will make the boiler so shiny it will be like a mirror. The only time when you should use emery paper on the boiler is at the bottom of the boiler: where all the soot collects from burning the fuel. You can use it here because 1) Brass polish alone won't be coarse enough to take the soot off and 2) It is an area of the boiler which cannot be seen on stationary engines and is hard to see on traction engines.

- Try not to get brass polish inside the threads. It is much easier to clean the boiler when all the threaded components such as the water level plug, safety valve and whistle are removed. However, in doing this, you are exposing the inside of the boiler. Ideally, the inside of the boiler needs to stay as clean as possible to maintain the performance of the engine: any rubbish such as brass polish that goes into the boiler will need to be removed by washing it out of the boiler with water.

- It will tarnish! The chances are that if the Mamod engine you are restoring is in its original condition, the boiler will be very tarnished and it will take at least a good few hours to clean. Once you have completely cleaned the boiler, it will start tarnishing again since the new brass layer is exposed to the oxygen in the air (so the top layer of brass starts oxidising). Within days, you will notice a change in the colour of the brass that it is not as bright or shiny anymore - do not worry! You don't have to re-polish the boiler as if you do this, you will be forced to keep doing it every few days. The boiler will naturally tarnish over time – this means the tint the boiler is at will darken but its shininess will stay constant.

Restoring a Mamod SE

The Mamod SE range are the perfect engines to restore if you are a beginner in restoration. Although the structure of the Mamod engines are fairly similar, the number of parts on the SE1/2/3 range is less than other engines such as some of the SP and traction engines.

You will also find that the SE range is one of the most collectable types of Mamod engines that were made. They have a lot of brass parts compared to other engines such as the TE1a green boiler traction engine which makes the engine look much better. As well as this, the SE range will always have a place in Mamod's history considering it was the first type of engine Geoffrey Malins started building for Hobbies all the way back in 1936 and is the only type of engine Mamod created that went through the whole of World War II.

As well as this, these engines are pretty cheap to purchase too. You can find on the internet that these engines range in between £20-£70 depending on condition (however, Mamods have become *extremely* collectable which causes their prices to continually rocket).

When it comes to buying an SE engine to restore, you need to make sure of the following things:

- Does the engine work? An engine that doesn't work will add lots more problems to your restoration. For someone with little experience with steam engines, only ever buy an engine which the seller has guaranteed can run.

- Are all the parts there? Although it is not the end of the world if the engine is incomplete, parts for Mamod engines are generally expensive and it will be much cheaper for you to buy a complete engine.

- How bad a condition is it? If there is rust on the engine, you need to look at how bad the rust has affected the parts on the engine. If the engine has been left for a long time, there is a chance the rust has become catastrophic which will result in you having to replace these parts as they are beyond restoration.

What You Need To Do First

When it comes to restoring your engine, it is a *must* to take several pictures of your engine before you do anything to it. This is because:

- You need to make sure you have 'before' and 'after' pictures for your engine.

- So that if any parts are missing, you can replace them.

- When you have finished restoring your engine and put it back together, you can use the original pictures of the engine before the restoration to compare with your newly restored engine as a secondary way to make sure you have rebuilt the engine correctly.

After that, you can then go about taking the engine part. You always need to take the whole engine apart and store all the loose parts in a transparent sealable bag. This is helpful because you won't lose any of the small parts and you can use a new transparent sealable bag to store the parts that have been refurbished.

Taking apart the SE model is pretty simple. The chrome on the boiler is held down by two screws underneath the base: the whole of the engine is held together by screws except the engine bracket. The engine bracket is held down by 5 rivets (which are actually hollow rivets as they are hollow in the middle). To successfully remove these, you will have to drill them out carefully. Make sure you choose a drill size that isn't going to expand the rivet hole. The firebox is kept on the base by means of screws and nuts that go through the boiler band that goes around the boiler.

By now, you should have all the small parts in a transparent sealable bag and the main larger parts next to it. The next step I always do when it comes to restoring my engines is to repaint the painted parts. The reason for this is because once I have finished painting the paints, they will need time to dry and the paint to harden. In that time, I could be restoring the smaller parts of the engine. Use your time efficiently!

If you don't know how to repaint a surface, have a look under the start of the chapter 'Universal Restoration Tips' (see page 67). You need to decide if the paint needs stripping or not (is there rust under the surface?). If there is lots of rust, it is best to strip the paint, put rust converter onto the surface, give it a day, smooth the surface off and then wipe with white spirit and re-spray. Usually, parts only need 3-4 coats of paint. More information on repainting a surface can be found under 'Universal Restoration Tips' (see page 67).

With the SE models, there are only three colours that are used on the painted parts:

- Red for the base and flywheel and the inside of the chimney. I found the best red paint is 'Ford Rosso' Red. You shouldn't worry too much about which red you buy: normal enamel red spray paint will work just as well because if the base and flywheel are painted the same shade of red, nobody will know the difference between the red you used and the original Mamod red.

- The firebox uses heat resistant satin black spray paint. However, this is difficult to find online let alone in stores. I did manage to find it but the price was around two times more than for heat resistant matte black spray paint. Therefore, if you are on a tight budget, go for matt heat resistant black paint. If not, you can go for satin heat resistant black paint. It is important that the paint is heat resistant since the firebox *will* get hot. If you do not choose heat resistant paint, the paint will bubble off the firebox, soften and ruin the look of the engine.

- The engine bracket uses green with a hammered finish (it has a bumpy finish). For best results, use a light green shade of Hammerite paint. However, every time I have tried to find this paint, it is not in spray form. Therefore, I resort to using Ford Highland Green (spray paint). This is the closest shade of green to the original Mamod green. The later models have a darker green finish close to Apple Green. However, I like to stick to using the lighter Highland Green as it makes the engine seem more vintage.

Important Tip!

If two parts do not fit, there is a reason they do not fit. Brute force does not solve the problem but will merely damage your engine parts.

If you are in a rush for your painted parts to dry, you can speed up the drying process by 'cooking' your parts in an oven. This might sound extreme but it is a process many companies use to speed up the process of paint drying. Make sure your part is on a tray that you do not really mind getting dirty and set the oven to a low/medium heat. 30-60 minutes of this will work wonders in hardening the paint.

If you are reading this, you should have taken the engine apart and left the painted parts to dry in a non-dusty room. For best results, leave the painted parts for 2-3 weeks to allow the paint to harden. However, if you are in a rush with your restoration, 4-6 days is okay but be careful with handling the paintwork making sure you do not bump or scratch it (be delicate with the painted parts in general as they will still be slightly soft).

The next step to the restoration is the cleaning of the metal parts. As listed under the chapter 'Universal Restoration Tips' (see page 67), there are four metals which you will be cleaning being brass, steel, chrome and copper.

Once you have cleaned your parts, put them in a new transparent sealable bag so that they are easily identified as having been cleaned and will not get dirty from the other uncleansed parts. The next step is to reassemble the parts since they have now all been restored.

If you are reassembling while the paintwork on the painted parts is still soft, you need to take extra care when putting the parts back together: especially when it comes to using pop rivets to connect the base and engine bracket together. On the original Mamod SE, they have used hollow rivets which are much more difficult to put in than normal rivets (but produce the best results as hollow rivets look more authentic than pop rivets). I tend to stick with normal pop rivets since it makes the engine look better with the aluminium effect but it is optional which you choose to use. After you have the engine bracket connected to the base, every other part can be put on in no particular order.

Below are some tips when it comes to rebuilding all of your restored parts to your engine and putting them all back together:

- Use the pictures you took before the restoration as a guidance to rebuilding your engine.

- The metal parts to the engine scratch very easily since they are polished and the scratches *will* show. Be extra careful with your tools around the boiler, piston, piping etc. as they are likely to scratch the engine if they are to come into contact with them.

- Don't *ever* force two parts together if they do not fit! Even if you know they should fit but the two parts don't (such as the flywheel and the crankshaft), do not force them to fit! The chances are that there is something impeding the movement of the parts together. Take your time and analyse the situation. For example, I find that there is some paint usually in the middle of the flywheel. This makes the crankshaft very tough to put through the hole and, to counteract this, I quickly go through, with a drill, the hole to remove all the paint from inside the flywheel.

- Make sure every single part is put back onto the engine. This is the reason you have put all your parts into transparent sealable bags. It is vital to the engine working that every part, including all screws, are put back onto the engine in their original location.

After you think you have restored the engine, look at the engine and compare it to the original pictures of your engine. Do they look the same? Once they do, it is then time to test run your engine. I tend to run the engine on compressed air first as the paint is still usually too soft to actually steam the engine up. I tend to wait around 2-3 weeks before steaming up after restoration just to make sure that the paint has hardened and will be durable when put under heat constraints.

Restoring a Mamod SP

The SP range was introduced by Mamod in the late 1970s to replace the successful SE range. Therefore, when it comes to the design of the engines, they are in fact very similar. However, there are key differences between the SE and SP range:

- The SP range has chrome surrounding the boiler supported by a black firebox surrounding.

- There is no exhaust piping on the SP1/2 models. The chimney is merely for appearance. The exhaust piping is only on the SP4/8.

- They do not use water level plugs anymore. Instead, the whole SP range uses the sight glass.

Since these engines are a bit more modern compared to the SE range and are still being built today, the prices of these engines tends to be a bit higher than that of the equivalent SE range. However, with this increase in price comes a better condition engine. At maximum, the engine cannot be more than around 30 years old. Therefore, you can be assured that the engine has not been used as much as an SE might have been.

When it comes to buying an SP engine to restore, you need to make sure of the following things:

- Are all the parts present? Many SP engines sold tend to have the chimney missing since they are only a decorative piece. However, the chimney online will cost you at least £10 to replace. Therefore, to save yourself the time and money of buying a replacement chimney, make sure all the parts come with the engine.

- Does the engine work? An engine that doesn't work will add lots more problems to your restoration. Only ever buy an engine which the seller has guaranteed can run.

- What is the condition of the engine? If there is rust on the engine, you need to look at how bad the rust has affected the parts on the engine. For the SP range, the chrome work is where you need to look most. If there is rust on the chrome work, you can remove this, if it is light rust, by using polish and cloth as you must not use emery paper on chrome work since this will remove the plating. The chrome work's condition is the most important part when it comes to trying to restore an SP engine to mint condition, as it covers most of the engine.

What You Need To Do First

I cannot stress enough that when it comes to restoring your engine, you always need to take the whole engine apart and store all the loose parts in a transparent sealable bag. This is helpful because you won't lose any of the small parts and you can use a new transparent sealable bag to store the parts that have been refurbished.

Taking apart an SP model is slightly more challenging than that of the SE range. First thing you should do is take apart the small parts such as the flywheel, piston, cylinder, crankshaft, whistle and safety valve. The engine bracket/s on the SP range are blue in colour and can be dismantled by drilling through the rivets holding it onto the grey platform or the main red base.

Fortunately, no screws or rivets are used around the boiler area: the firebox is kept attached to the base by clipping it into two holes in the base. After that, the chrome work can be easily removed as it is held in place by being clamped by the firebox. The main problem occurs when trying to remove the boiler from the firebox. The boiler is held to the chimney-end of firebox with two copper rivets. Whenever I do a restoration, I always try my hardest to not tinker with the boiler. You can still easily clean the boiler and firebox together without removing the rivets. However, if you do decide to drill out the rivets, make sure you use copper rivets as these will withstand the pressure and heat from the boiler. On the SP1/2 engines, the engine bracket is mounted to the top of the boiler. Again, it is best to leave this as if you remove it, you are creating more problems for yourself if there is a gas leak after assembling the engine together again.

By now, you should have all the small parts in a transparent sealable bag and the main larger parts next to it. If you feel like you are going to lose the larger parts, you can store them in the box you bought your engine in. The next step I always tackle when it comes to restoring my engines is to repaint the painted parts. The reason for this because then once I have finished painting the paints, they will need time to dry and in that time, I could be restoring the smaller parts of the engine. Use your time efficiently!

If you don't know how to repaint a surface, have a look under the chapter 'Universal Tips' (see page 67). Basically, you need to decide if the paint needs stripping or not (is there rust under the surface?). If there is lots of rust, it is best to strip the paint, put rust converter onto the surface, give it a day, smooth off and then wipe the surface with white spirit and respray. Usually, parts only need 3-4 layers of paint. More information on repainting a surface can be found under 'Universal Tips' (see page 67).

Colours Used

With the SP models, there are only three colours that are used on the painted parts:

- Red for the base and flywheel. This best red paint is 'Ford Rosso' Red. You shouldn't worry too much about which red you buy: normal enamel red spray paint will work just as well because if the base and flywheel are painted the same shade of red, nobody will know the difference between the red you used and the original Mamod red.

- The firebox chimney and engine bracket (which only applies to SP1/2 models) uses heat resistant satin black spray paint just like the SE models. The chimney for some of the models that doesn't have an actual functioning chimney can use non-heat resistant paint since it is not going to be in contact with anything hot. However, it is best to stick to using the same spray can for painting parts of the same colour (don't overcomplicate the restoration using different types of the same colour paint!). The best way to paint the engine bracket on the SP1/2 models is to polish the boiler first and then mask everything up (using masking tape) except the part you want to paint.

- The engine bracket on the other SP models uses a dark blue smooth paint. The below picture shows one of my restored SP4 models, I decided to stick to the same green spray paint used on the SE models because I think it goes better with the red and black of the base, chimney and firebox. However, if you want to be accurate and use blue, it really does not matter which paint you use. Any dark blue spray paint will do when it comes to spraying the engine bracket. If you want the closest shade to Mamod's blue, I would recommend you go to your local DIY store and compare the shade of blue on the engine bracket with the shades of blue in store.

- The platform which holds the engine bracket on a SP4 does not have a specific colour. For this reason, all you need to make sure is that you buy a matt light grey spray paint.

You may think I am being too general with the paint colours but when it comes to bright contrasting colours such as blue, grey and red, subtle differences in shade from the original are hardly noticeable.

For this SP4 engine, I chose a green coloured engine bracket similar to the SE range out of preference.

If you are reading this, you should have taken the engine apart and left the painted parts to dry in a non-dusty room. For best results, leave the painted parts for 2-3 weeks to ensure the layers of paint harden. However, if you are in a rush with your restoration, 4-6 days is okay but be careful with handling the paintwork making sure you do not bump or scratch it (be delicate with the painted parts in general as they will still be slightly soft).

The next step to the restoration is the cleaning of the metal parts. As listed under the chapter 'Universal Tips' (see page 67), there are four metals which you will be cleaning being brass, steel (or Mazak), chrome and copper.

Once you have cleaned your parts, put them in a new transparent sealable bag so that they are easily distinguishable that they have been cleaned and so that they will not get dirty from the other uncleansed parts. The next step is to reassemble the parts since they have all now been restored.

If you are reassembling while the paintwork on the painted parts is still soft, you need to take extra care when putting the parts back together: especially when it comes to using pop rivets to connect the grey platform and engine bracket on SP4 models and above together. Fortunately for restorers, Mamod chose to use normal pop rivets for the SP range. After you have the engine bracket connected to the base and the firebox, boiler and chrome housing clipped into the base, every other part can be put on in no particular order.

After the engine is complete, compare your engine to its original pictures to make sure everything is where it should be and then try to steam it up (preferably, at first, with pressurised air). If it does not run, you can find out why in the 'Problems with Starting Your Engine' chapter (see page 98).

Restoring a Mamod Minor

For anyone who hasn't restored an engine before, I would highly recommend you try restoring a Mamod minor engine first as they are by far the easiest and most simplistic engine Mamod have ever made. The Mamod minor range only has two models: the MM1 and MM2.

- The MM1 is the oldest version of the minor range and features the smallest boiler made by Mamod.

- The MM2 was introduced after the MM1 as a 'big brother' to the MM1 with a larger boiler.

These engines are the smallest Mamod have ever made and probably will ever make. This makes them highly collectable by collectors and restorers as are they relatively cheap to purchase and easy to restore. Here are the parts of the MM1 and MM2 engines listed and how you can go about restoring them:

- Boiler. The boiler can be removed by unscrewing the two screws that keep the chrome band wrapped around the boiler and onto the base (this will ultimately bring the whole engine apart). Once you have the boiler by itself, you will find that the engine bracket (which holds the crankshaft) is connected to the boiler itself. Therefore, when it comes to restoring the engine, *restore the boiler first*. Although I have said for other engines to do all the spray painting first, if you spray paint the engine bracket and then try to polish the engine, you will find it difficult to not interfere with the paintwork on the engine bracket. Polish the boiler first and then cover the whole boiler in masking tape so that when you do paint the engine bracket, only the engine bracket will be painted.

- Chrome boiler band. The chrome boiler band is a feature on all of the minor range just like it is on the SE range. To restore this, polish gently if you can see the chrome. If there is rust on top of the chrome, gentle remove it with extremely fine emery paper. Remember that the aim is to not go through the thin layer of chrome.

- Firebox. The firebox is similar in design to that of the SE range. It is made from steel and coated in satin heat resistant black paint. If the paintwork is in good condition, you won't need to restore the firebox (you can simply clean it of dirt). If you find there is rust on it or cracks and lines through the paintwork (this means there is rust forming *under* the paintwork), you will need to remove the paintwork down to bare metal, treat the rust with rust converter and re-spray. As I have already said with the colour for fireboxes, satin black heat resistant paint is hard to come by (although it is becoming progressively cheaper). Therefore, it is best, if you are on a tight budget, to just spray the firebox with matt black heat resistant paint. After this, you can then purchase a Mamod logo online to put back on the firebox.

- Flywheel. I always find that the flywheel is one of the easiest parts on any engine to restore. There are two jobs you need to do with the flywheel: 1) clean the metal around the outside so it's shiny and 2) re-spray the red paint on it. Always clean the metal around the outside of the flywheel first because if you spray the inside of the flywheel first, you may damage the paintwork when it comes to cleaning the outside metal. Fortunate as it is, I have a pillar drill which I mount the flywheel into, rotate it and apply emery paper with increasing grit levels so the surface gets smoother and shinier. The last step is to use a piece of old towel with metal polish and apply lightly to the flywheel. However, if you do not have a pillar drill, you can still polish it using just an ordinary drill. I attach the flywheel to the crankshaft, make sure it is secured to the crankshaft tightly and then insert the crankshaft into the tip of a drill and tighten (don't worry, the crankshaft doesn't scratch and if it does, it can be easily smoothed out with emery paper). You can then rotate and apply, ideally, a sanding sponge to the outside of the flywheel (and the inner side too). You can then mask the area you don't want painting so that only the inside of the flywheel is painted.

⬆ Here is the how the flywheel connected to an ordinary drill looks like

- Base. The base can be restored by re-spraying the surface with red spray paint. If you look under 'Universal Tips', you will find what to do if you have rust or not on the base and how you should respray the base. If you do not want to, it is not essential to re-spray the underside of the base. However, you will find a slight difference in shade and may want to get rid of this by spraying the underside too.

The Mamod Minor range are very simple engines to restore and put back together. However, like always, remember to put all the parts into a transparent, labelled resealable bag so you do not lose any screws or small parts.

The only real problem with these engines (as well as the SE3) is that trying to find an end cap for the smaller sized (for the SE3, larger sized) boiler is difficult to find. Therefore, if you choose to restore a Mamod Minor or a SE3, make sure that the boiler is in perfect working order otherwise you will find yourself in a bit of a pickle trying to find boiler parts.

Advertisement:

Made in England since 1937

Order on-line direct
www.mamod.co.uk

Centurion
1313C
Working Steam Tractor
Double Action Piston Cylinder

NEW
Double Action Piston

The Centurion incorporates the successful design of the new double action piston valve cylinder.

- Silver solder construction
- Re-heating coil
- 9.5 bore
- 14.0 stroke
- Displacement lubricator
- Forward and reverse actikon by slip eccentric

Specification:
Dimensions of model: 178 x 137 x 270mm
Gross weight: 2100g

Mamod Limited
Unit 1A Summit Crescent Industrial Estate Smethwick Warley West Midlands B66 1BT
T: 0121 500 6433 Fax: 0121 500 6309 E: accounts@mamod.co.uk www.mamod.co.uk

Restoring a Mamod Traction

For this chapter, I will mainly look at how to restore the TE1/a traction engine, SR1 Steam Roller with little bits about the SW1 Steam Wagon too. Traction engines are the hardest engines to restore due to the complex parts and different colour schemes used. For example, with the SE range, there are only three colours ever used. With traction engines, there are five different types of colours used. Before I look individually into each engine, below is a list of paints used by the Mamod traction engines and where they apply to on the engine:

> When it comes to painting your Mamod engines, if there are parts of the engine that are the same colour, make sure you use the same shade of paint. For example, if the flywheel and wheels are red, use the same shade on both parts.

- Green. This is used on the boilers of the majority of traction engines. For early engines (around and before late 1970s), the closest shade is Apple Green. The later engines then went on to use Dark Apple Green. The colour that is closest to current Mamod traction engines is either Dark Apple Green or British Racing Green. Since this is on the boiler, you will need heat resistant paint. The paint, if not heat resistant, will struggle with the temperature the boiler reaches and may bubble or soften. Therefore, if you do not use heat resistant green paint, do not touch the paint work after steaming up as it may still be soft and produce marks.

- Red. The red is used on some of the models flywheels and wheels. The correct shade of red for this is the same as the stationary engines: Ford Rosso Red.

- Black. This is used on the firebox and chimney (and other parts) of the traction engines. The correct paint for this is heat resistant black satin.

- White. White is used on the canopy of some of the engines. Fortunately, the paint used can be just normal white spray paint. It is wise to make sure it is enamel as this will make it much harder than normal paint and help the canopy's coat survive any knocks and bumps.

- Decal. On some of the engines, decals are used to improve the appearance of parts such as the side of the wagon on the SW1 or the canopy on tractions engines.

Decal for the sides of the SW1 Wagon →

- Blue. Unfortunately, I haven't got a shade of blue accurate enough to be close to the blue Mamod used. The best thing to do for this is take your engine and compare shades at your local DIY store. If not, I would go for a dark blue shade. However, with bright colours, such as blue, the shade of blue you choose will not matter as much as whatever shade of blue you get will still contrast well to the other colours.

Now that the paint colours have been sorted, I will look into each of the three engines into a bit more detail on how to restore them. The first step to restoring your traction engine is knowing just *exactly* what type of traction engine you have (look at a the 'History of Mamod' and 'The Range of Mamod Models' section to better identify your traction engine):

- If your traction engine does not have a lever to control the speed and direction of the engine but a regulator by the pipe's entry to the boiler, this is an early TE1/SR1. These engines are much rarer and worth more.

- If the boiler to your traction engine is unusually light green, this is also an early traction engine ranging from before the late 1970s. If it is darker, it is a more recent engine. Another key way to tell the age of an engine is through its condition. Earlier engines will tend to have more rust formed on them than newer engines due to the many more years of service they have endured.

- The box. If you are lucky enough to have a box come with your traction engine, this will help significantly in identifying the age of the engine. To best identify the engine's age from the box it came in, I would recommend using www.mikes-steam-engines.co.uk. It has a whole selection of Mamod engines from over the years with boxes. You are sure to find your engine there.

Now that you know what engine you have, you can find the right paint to suit it (right shade of green). It is only the boiler colour that changes over the years so you won't have to worry about the other colours the traction engines have: they will stay the same shade no matter what year the engine was produced.

Like with all restorations, the first step is to dismantle the engine and put the small parts in a transparent sealable bag. The engine bracket, piston/cylinder assembly, flywheel, whistle, water level plug and safety valve can be easily removed. However, to remove the rest of the engine takes a bit more time:

- To take off the wheels, you will need to remove the hubcaps. Hubcaps have tangs acting a bit like barbed hooks for those that fish: you push it on and it is then hard to get off. Hopefully the following diagram on the next page will best describe this:

This tip applies to all of the traction engines.

(Since all of Mamod's mobile engines have wheels on axles which require hubcaps to keep the wheels on the axle)

(Continues onto next page)

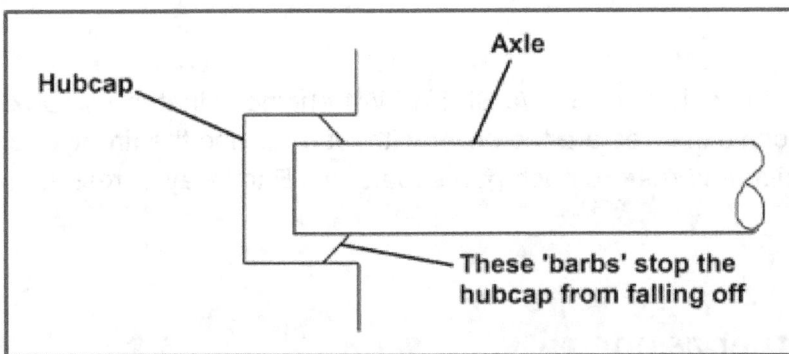

This makes clear that if you attempt to pull the hubcap straight off, you will damage or snap the barbs causing the hubcap to become worthless (as it won't stay on when you next put it back together). To take the hubcap off, you need to use some pliers preferably with something soft or rubbery on the inside (to save scratching the hubcap) and slowly rotate the hubcap and pull. Through doing this with patience, the hubcap will eventually come off and will still have the barbs/tangs intact. Once the hubcaps are released, you can then remove the wheels and back axle.

As you can see above, there are two types of hubcaps your Mamod engines may possess. Your engine will either use the chrome hubcap, together with barbs/tangs (as on the top left diagram), or your engine will have plastic black hubcaps (which are easier to remove, since the plastic tangs can be bent further that the chrome hubcap's tangs).

- The chromebox and firebox are mounted onto the end of the boiler. For successful removal, you will have undo the rivets on top of the chromebox and at the end of the boiler. The rivets into the boiler are made out of copper in order to cope with the changing pressures and temperatures of the boiler. If you remove them (through drilling them out), make sure you use copper pop rivets when reassembling the engine again.

- It is preferable not to remove the front axle when restoring the engine. Since the front axle is steel, it can be easily cleaned whilst remaining connected to the front of the chimney/boiler area. Remember to try to keep your restoration as simple as possible.

At this juncture, you should have dismantled your engine into different parts. For the small parts of the engine, you can restore them quite easily (look at the chapter 'Universal Restoration Tips' for restoring different materials on page 69). The main problem comes with restoring the boiler section: especially if you chose to keep the firebox and chromebox connected to the boiler.

> Fortunately, many of the parts on the traction engine are similar to that on the stationary engines. Therefore, my advice would be to look over the stationary section and Universal Restoration Tips for restoring the other parts of a traction engine. If you are unable to find the part, have a look in the Glossary section at the back of this book for guidance.

Restoring a Traction Engine Boiler

The main problem associated with restoring the boiler of a TE1/a/SR1/a/SW1 engine is that the firebox is connected to the boiler. This means you can either restore the boiler without removing the firebox from the boiler or remove the firebox from the boiler and restore each part separately. Each way of restoring the boiler has its pros and cons:

Keep the boiler and firebox together as one part

+ The firebox is connected to the boiler via copper pop rivets (since copper is a good material to use for rivets when under pressure and heat). Keeping the parts together saves you the hassle of taking them out and putting new rivets in, reducing the chances of problems occurring.

+ It will save you a bit of money since copper rivets are in the region of £2-£4 to buy.

+ In general, the less parts you have to remove, the quicker the restoration will be and the more chance there is that parts remain intact, allowing you to easily put them back together again.

- If the boiler and firebox are still together, you are going to find it a real struggle cleaning the section of boiler inside the firebox. Considering this is the area that has got the most soot, it is not ideal. However, it is hardly visible on the outside so not that big a problem.

- The biggest problems arise when it is time to respray the parts with paint. You will have to use mask and protect the boiler when painting the firebox and vice versa. Since you will also have to wait for the paint on one part to harden before you paint the other (as you will need to protect that coat against being contaminated with any other paintwork you are undertaking), it can take quite a long time to paint both parts.

What I decided to do when I restored a boiler and firebox assembly is to respray the firebox with heat resistant black satin paint but to not respray the boiler. Trying to spray an assembly with two different colours is extremely difficult and so it was far easier to strip the green coat of paint off the boiler and convert the engine into a brass boiler steam engine (which I think also looks much nicer than the traditional green boiler engines).

Separate the boiler and firebox

+ Painting the parts will be *much* easier. As well as this, you will be able to fully clean the boiler (the section of boiler that is sootiest is inside the firebox) and respray the whole boiler: not just a section of it.

+ Separating the parts makes it much easier to clean them, such as stripping paint, removing dust from surface of parts with white spirit and so on.

- The main problem occurs with how the firebox and boiler are connected. Depending on what engine you have, you will either have a water level plug or a sight glass. Either way, there are two copper rivets that connect the firebox to the boiler. I tend to try to leave the boiler alone as much as possible, as this is the main part of the engine. However, if you want to separate the boiler from the firebox, you will have to carefully drill out both the copper rivets.

It is important that when you assemble the boiler and firebox back together you use copper rivets again. Aluminium rivets cannot stand the pressure and heat which the copper ones can. This is a safety warning since the pressure inside the boiler can cause the rivets to fly out at tremendous speed.

Which option you opt for from the above depends on your engine's condition. If the boiler's paint finish is in really good condition, leave it and don't dismantle the boiler/firebox assembly. If the paint is in bad shape and you want to convert it to a brass boiler engine, then you can manage this without any dismantling. For this reason, only dismantle the boiler and firebox from each other if you really have to.

When you are restoring the boiler of your traction engine, remember to mask up the areas of the boiler you don't want to spray (or want to spray a different colour). When you are spraying the boiler, you will need to cover up the whole chimney and front axle section as well as the steel band just behind the chimney.

As well as this, when it comes to spraying the chimney/front section of the boiler assembly, I tend to spray that whole section with heat resistant black satin paint. However, since the temperature of this section is going to be a bit lower than that of the boiler, you might get away with paint that isn't heat resistant (although, to stay safe, I advise you to use the same heat resistant paint you used on the firebox). To make the Mamod raised emblem visible, once again, at the very front of the engine, use a wooden block and emery paper and rub against the front of the engine. This will take the paint off at the front revealing the Mamod logo and engine name for some engines.

Restoring the other Parts of a Traction Engine

Here are two other parts found on the traction engine and how to restore them:

- Wheels. Treat the wheels like the flywheel on the SE/SP/Minor range. Clean the outer side of the wheel first with emery paper and metal polish and then strip the paint and then re-spray the inside of the wheel with the outer side of the wheel having already been masked off with masking tape. Depending on what engine you're restoring will depend on what colour to paint the wheels (red for most traction engines, black for the brass boiler engine yellow for the Showman special and tractor kit).

- Chromebox. Just like with the SE/Minor chrome band and SP chrome housing, be careful not to scratch the chrome off since the layer of chrome on the chromebox is very thin. If rust is visible, try and remove the rust with light emery paper (better yet, replace it since it will get progressively worse). If it doesn't have rust, use metal polish and a rag to lightly clean the chrome. If it doesn't get shinier, don't over rub, as this will only cause the chrome to become duller in colour, as you rub through the chrome to the steel layer beneath.

Restoring a Steam Roller Engine

I have to admit that out of the traction engine and steam roller, I do prefer the steam roller based on its look. For us restorers anyway, the engines are nearly identical with what parts they use which means many, if not all, of the tips for restoring a traction engine applies when it comes to restoring a steam roller. For this reason, have a look at the 'Restoring a TE Traction Engine' section if you are restoring a steam roller (see page 87).

The only real significant difference between the traction engine and steam roller are the rollers (instead of two wheels on the front axle). When it comes to restoring the rollers, treat them like the traction engine wheels, as that is what they are but extruded.

When it comes to selecting what colour to choose to spray your engine, again, think about what the original colour was and how easy it is to spray the parts. If you are going to leave the firebox attached to the boiler, then you may opt to create a hybrid brass boiler steam roller (it will be the first of its kind)! However, a word of warning, Mamod steam rollers have ever only had one colour scheme that being a green boiler, black firebox and chimney with red wheels (unlike the traction engine that has a few more colour schemes). Therefore, it will made more apparent to others you have restored your steam roller if you change the colour scheme to it (which you may want to do to distinguish your engine).

Advertisement:

Made in England since 1937

Order on-line direct
www.mamod.co.uk

SR1A
Working Steam Roller
C89

A superbly realistic model of an early road roller.
Also available in kit form (c/w canopy).

Specification:
Pack Dimension: 285 x 150 x 185mm
Gross weight: 1940g

Mamod Limited
Unit 1A Summit Crescent Industrial Estate Smethwick Warley West Midlands B66 1BT
T: 0121 500 6433 Fax: 0121 500 6309 E: accounts@mamod.co.uk www.mamod.co.uk

Restoring a Canopy

As you might be aware, both the traction engine and some steam rollers come with canopies that covers the steam engine and protects it from rain. Since the canopies are such a different part compared to the rest of the parts in a traction engine or steam roller, it deserves its own section on how to restore them.

The canopy is a decorative piece that does look hard to restore. However, in essence, it is extremely easy: what makes it look difficult are the decals adorning it. Here is a step by step process to follow when restoring a canopy:

1. **Remove the canopy supports**. These are the two metal straps that connect to the back of the engine and support the weight of the canopy. Depending on what type of canopy you have, they are either riveted into the top of the canopy (which require drilling out the rivets) or held by screws.

2. **Strip *all* the paint**. I mention 'all' because you will find that most canopies will be in poor condition with rust present under the surface of the paint. As well as this, the white paint has probably stained over the years to off white (or Old English White). Therefore, if you choose to keep some of the good paintwork and then respray, you will see a noticeable difference in the shades of white paint. Use emery paper to get the majority of the paint off and then go over the canopy with rust converter and white spirit to make sure all the rust is treated and paint is removed.

3. **Respray the top of the canopy**. For this, use normal white enamel gloss paint. Have a look in the 'Universal Restoration Tips' chapter (see page 67) for more on respraying parts. You will need to take your time on the canopy and respray one side of the canopy first, let the paint dry and then respray the other side (the whole of the top canopy is white). It is also advisable to give the canopy a few more layers of paint than other parts of the engine since a scuff on black paint is not as noticeable as it is on white.

4. **Restore the canopy supports**. The canopy supports are made of steel and can be cleaned with metal polish (remember that emery paper will scratch the supports so try and stick to metal polish and a rag).

5. **Add decal to the white canopy**. Once the canopy has been completely resprayed, you can then go about adding the decorative pieces to finish the piece off. The great thing about Mamod canopies is that they can be customized in the sense that you can add whatever decal you want to the canopy. For some people, they like the canopy to be a plain colour. Others like different decals on the side to the original Mamod canopies – a good example of this is the Fred Dibnah decal you can purchase on many websites. For this reason, browse the internet and find the decal that you feel best suits your engine and canopy.

6. Reattach the canopy supports. Depending on what type of canopy you have, you will either have screws to reattach the canopy support or rivets (which you will need to buy new ones, as the previous rivets would have been destroyed disassembling the parts).

Considering the canopy is only cosmetic, there is always the option of leaving the canopy off your engine if that is what you prefer. I tend to find that TE1a engines look nicer without the canopy insomuch for the reason that I feel the canopy prevents the chromebox, whistle and piston/cylinder assembly from being clearly seen. However, that is just my preference!

Restoring a Lumber Wagon Trailer

The wagon trailer is a great restoration to do because they are extremely simple to restore compared to other Mamod parts and end up looking absolutely stunning at the end of the restoration process.

When it comes to restoring the wagon trailer, there are far fewer parts to worry about which are listed below:

- Four wheels. These wheels are near enough exactly the same as the wheels found on the front of the TE1/a traction engine which include the barbed hub caps (see pages 88-89).

- The connecting rod. This is the long rod that goes between the front and back of the wagon trailer.

- The front and back of the wagon trailer (parts in green).

With the wheels, we will restore them the same way we have restored any flywheel/wheel on a Mamod. You will want to polish the outer surfaces and strip the paint off before spraying the wheels:

1. Using very light emery paper (800-1200 roughness), rub the outside of the wheel (basically the parts which are polished in the above picture). To help aid getting off the paint, you can you emery paper for that too.

2. Polish the wheel with metal polish and cloth to get the wheel to shine.

Important Tip!

It is important to remove **ALL** of the paint from the surface of an about-to-be-sprayed surface, seeing that if any paint remains, the surface finish of the new paint layer will be bumpy due to the different layers of paint.

3. If all the paint is not removed from using emery paper, carefully use white spirit to take the remaining paint off.

4. Once all the paint is off the wheel, the next step is to prepare the surface for spray painting. Cover the areas of the wheel that are not getting painted with masking tape and prepare the areas that are being painted by rubbing with white spirit (white spirit removes all the dust from the surface and degreases the surface promoting good paint adhesion).

5. Spray the wheels from left to right (or right to left) with the spray canister about 30-50cm away. Make sure to spray smoothly and at a steady speed so that all of the wheel are sprayed with equal amounts of paint.

6. Leave the wheels in an open area away from dust for around ten minutes to let the paint dry and harden. After ten minutes, repeat step 5 and spray the wheels again. Do this at least three times until the red is completely vivid on the wheel and has at least three layers (it is better to have many thin coats rather than one thick coat).

The same process can be applied when it comes to spraying the front and back of the wagon trailer (this time, instead of red, spray with Apple Green. As well as this, remember to carefully take the hubcaps off - bearing in mind that they are barbed and forcing them off will result in the barbs/tangs breaking making them unusable when reapplying back onto the wheel axle.

Problems with Starting Your Engine

I will be lying if I was to say that every one of my restorations has gone to plan. The reality is that problems will always occur with every restoration: sometimes small and sometimes big. Therefore, here are the main problems you may encounter when restoring your Mamod Steam engine with potential solutions.

My engine does not start! Why does it not work? Help!

The reasons for an engine to not run are not always 'big' problems and can be easily fixed if diagnosed correctly. When it comes to diagnosing, although you can test run the engine the traditional way with water and fuel, it is far easier to use pressurised air from an air compressor (or, if on a budget, a foot pump works just as well) and it will also protect your engine against getting sooty and dirty.

When diagnosing the engine if it is not working, it is important to analyse all areas of the engine listed below to find the root cause of the problem.

Diagnosis #1 - The Boiler

The first diagnosis involves the boiler. The boiler is where the water is converted to steam to produce a pressure which works the piston/cylinder assembly. The only reason the boiler could be the problem for an engine not to work is because it is not completely gas tight. The reasons for this are because:

- There is a leak in the connection of the piping to the boiler. For most engines, the piping is soldered onto the boiler. Therefore, to make this gas tight, you can re-solder the joint and add a little more solder onto the area you are soldering (with a dab of flux). For other SE engines, they use a rubber O-ring pressurised by a threaded nut to create a seal between the boiler and piping. Considering the engines are likely to be tens of years old, the rubber O-rings may have perished and left an air gap.

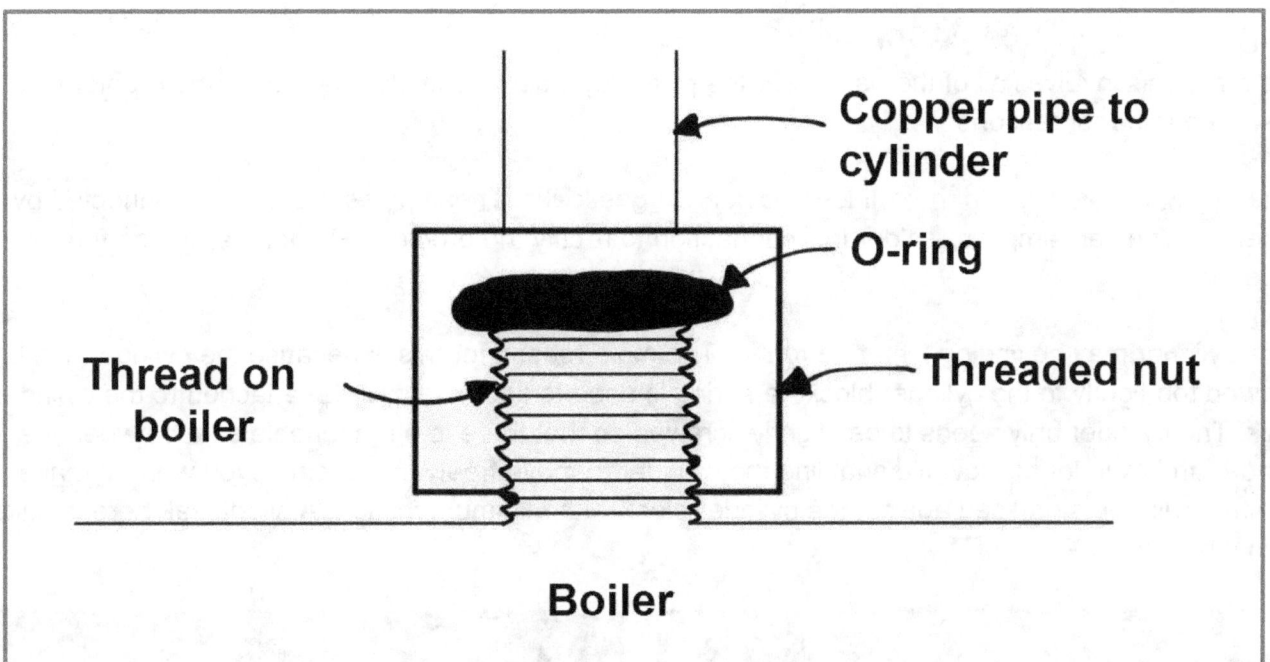

As you can see from the diagram, the gas tight seal is created by the threaded nut squeezing the O-ring over the copper pipe and onto the thread on the boiler. Without the O-ring, there is no gas tight seal and the boiler will lose pressure quickly. Therefore, make sure the pipe to your boiler is gas tight like the diagram above or soldered depending on what model SE you have.

- The washers are damaged. Let's face it, most Mamod engines are pretty old and, because of this, there is a strong possibility that the washers will have faded away over time, leaving no material to create an airlock between the boiler and the whistle, safety valve or water level plug. Therefore, it is always good to replace the washers on the engine if you are not happy about the condition of them or if you suspect/know they are not gas tight.

> Of course, if you have successfully restored your boiler using the chapter 'Universal Restoration Tips' (see page 72), you will have a boiler that is mechanically in perfect working order and you should not be having problems with your boiler. You can therefore eliminate the boiler as your problem and move onto the next diagnosis.

- There is a gas leak passed one of the threads (safety valve, whistle or water level plug). If the thread has stripped (meaning instead of screwing in, the safety valve/whistle/water level plug moves freely in the threaded insert), you will need to replace the threaded insert. This involves heating up the boiler to liquefy the solder around the insert. Take the threaded insert out, replace the threaded insert (which requires removing the end cap to the boiler to get the insert out and the new one back in) and solder it back in using the capillary action to suck the solder around the joint.

To diagnose a leak from the boiler, you can simply get the engine running on pressurised air from an air compressor or foot pump and either:

- Put your fingers or ear by each thread. If you feel or hear air escaping, you know this might be the reason for it not steaming up.

- Cover the suspect areas in soapy water so that if air is leaking, bubbles will appear.

Diagnosis #2 - The Piston/Cylinder Assembly

This diagnosis involves all of the parts near the piston/cylinder assembly. Therefore, the problem can be one of a number of reasons:

- The piping connections to the cylinder block have gas leaks. These pipes are always connected by soldering. You can simply re-solder this connection to rectify the problem. Remember to use flux when soldering.

- The cylinder may be finding it hard to rotate. The main reason for this is because the cylinder has been screwed too tightly to the cylinder block - a spring is used to keep the cylinder attached to the cylinder block. The cylinder only needs to be slightly screwed so that there is a reasonable seal between the cylinder and cylinder block while enabling the cylinder to move freely. In essence, you want to make sure the cylinder is pressed against the cylinder block with minimum friction. A slight leak here is quite normal behaviour.

Screwed in too tightly

Too much friction between cylinder and engine bracket makes it difficult for the cylinder to move and rotate

- The cylinder is finding it hard to rotate because of mechanical pressure caused from the piping slightly bending the engine bracket. This occurs when restorers have bent the copper piping when restoring it to a shape which is different to its original shape. This causes the piping, when reassembled, to exert a force (or mechanical pressure) on the engine bracket causing it to become out of alignment with the base. For the piston/cylinder assembly to work, it needs to be completely perpendicular (at right angles) to the base. For example, the engine bracket to the right would not allow the cylinder/piston to work since it is slightly bent.

If engine bracket is not straight, the cylinder/assembly will find it hard to move and rotate

- There is steam escaping between the piston and cylinder.
If this is the problem, you will need to buy a completely new cylinder and piston assembly. Saying this, air/steam will naturally escape through the cylinder anyway. Only replace this assembly if lots of steam escapes from the cylinder (and I mean lots)! You cannot do anything to make the cylinder and piston gas tight again, unfortunately. If you want to be extra safe, buy a completely new cylinder and piston assembly. If you know it is either just the cylinder or piston that is the problem, you can always risk trying to replace just one (although you cannot guarantee a good gas seal with the old part you are not changing which is why I recommend buying a new assembly).

Diagnosis #3 - The Crankshaft

The crankshaft has the potential to cause the engine to not work if the pin that connects the end of the piston to the crankshaft is restricting any movement from occurring. This transpires when the pin on the crankshaft is at the wrong angle or corroded (so is very lumpy). For this reason, it is pretty easy to diagnosis the crankshaft as the problem. For example, refer to the picture to the right. The pin in the crankshaft would stop the engine from working as it is corroded and lumpy and at an angle too (extreme version but you get the idea!).

Diagnosis #4 - Copper Piping

If I was to give you a tip, it would be to not mess around with the piping on your engine at all (although you would have heard this many times before by me in previous chapters). If you 'work' the piping by bending it, it will eventually snap. This is because you are creating dislocations between the atomic crystalline structure which, when located around a singular point (being the area that is being bent), will cause fracture to occur (this is called work hardening). Working the piping can also cause the gas path within the piping to decrease in size. Think of the piping as a hose pipe. If you bend the hose pipe, no water will flow out or the flow rate of the water will decrease. The same applies to the copper piping. If you suspect the copper piping to be the problem that the bore has collapsed, you will have to replace the piping.

Diagnosis #5 - Have You Steamed The Engine Up Correctly?

You may think your engine has some serious flaws from not starting up. However, the problem may be much simpler. If you have used an air pressure unit to start your engine up, have you used a high enough pressure to actually get the engine moving? For example, to get an engine to work on a foot pump requires somebody to constantly use the foot pump at quite a fast pace (around 1-2 presses on the foot pump per second).

Similarly, if you have steamed up the traditional way with water and fuel producing steam, have you put enough water into the boiler and fuel under it? Typically, an engine works best where the boiler is around 2/3 to 3/4 filled with water. Too much water and no pressure from steam can be produced and the feed pipes may be initially blocked with water. Too little, and there is not enough water to produce steam to create a high enough pressure (which will result in boiler damage).

Another common problem is with trying to start a traction engine with the drive band attached to the back wheels. If you are lucky, it should be able to run on the floor like normal. However, through my experiences with mobile engines, I have always found that to actually make the traction engine move forward, I needed to steam the engine up without the engine's back wheels touching the floor, get it going at maximum speed and *then* drop the engine onto the floor. The speed the flywheel is then rotating at will make it harder for the engine to stall when the back wheel connected to the flywheel hits the floor.

Ultimately, out of all of the above diagnoses, for a non-working engine, the most common are diagnosis #1 and #2. If you have bought an engine that has been neglected for years, it is bound to have problems with its soldered joints, washers and threads. Diagnosis #2 can be avoided if you do not change the shape of the piping. However, if you do change the shape of your piping, a reason for your engine not starting could be #2. If you are still unsure about how to steam up, check the chapter 'Steaming Your Engine Up' (see page 32).

Buying Parts for Your Engine

> This section will highlight the main areas you can source parts for engine online.

Hopefully, you won't have to read this section too much since buying parts for your engine will cost you money! However, with restoring anything, there could be parts missing or if the parts are there, they may be damaged. Therefore, you are most likely going to have to buy parts for your engine sooner or later.

I would recommend that when you go to buy parts to look at three or four of the websites below, put the parts you want into a basket and see which is cheapest, insomuch as they are all generally the same price plus or a minus a few pounds.

Tip

The biggest cost factor for buying parts online is the postage. Therefore, make sure you know **every** single part you need to buy to restore your engine. I would heavily recommend you **buy in bulk** to save on postage. If you have four or five parts to buy, you will be saving roughly £10 alone on postage.

MamodParts.com or ManorModels.com

Pros
+ Large range of Mamod parts on the internet.
+ Reasonably priced.
+ Dedicated 'Enhancements' section.

Cons
- Some of the parts go out of stock quickly and take a long time to get replenished.

MamodParts.com should be one of the first places you look for Mamod parts. They have the largest range meaning you are likely to find whatever part you are looking for.

Forest-Classics.co.uk

Pros
+ Large range of Mamod parts.
+ Usually hold good stock of parts.

Cons
- Pricing of some of the spares can be slightly expensive.
- Website can be slightly confusing to use and navigate.

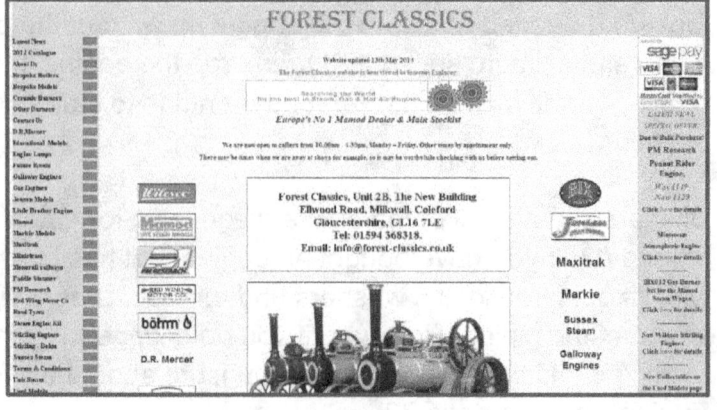

Forest Classics is another essential website to look at for spares. I tend to find that you are more likely to find what you want with Forest Classics than with Mamod Parts/Manor Models: the only difference is that you will probably pay a little more, which, if Mamod Parts/Manor Models doesn't have in stock, is worth the extra money to save time.

ModelEnthusiasts.com

Pros

+ A nice designed website to navigate around.
+ Supplies Mamod parts, plus extra accessories, to 'spice' up the look of your engine.

Cons

- Some of the parts have a premium price.
- Range of parts for sale are not as extensive as those of Manor Models and Forest Classics.

Model Enthusiasts is a good website to use. However, some of the parts are, unfortunately, slightly more expensive than Manor Models and Forest Classics. However, this doesn't mean *all* the parts are more expensive: some are actually much cheaper. This is why you will need to surf the web to see which website, overall, suits you best for all your parts, including postage.

Preloved.com

Pros

+ Prices of engines can be ridiculously cheap since people selling them sometimes underestimate the true value of their engine.

Cons

- Preloved.com is a free membership. Nonetheless, the full membership to see adverts as soon as they are listed costs £5/year which is something worth investing in (free, you have to wait 10 days after posting to contact seller).
- It is unpredictable how many adverts on Mamod engines will be displayed at that time.

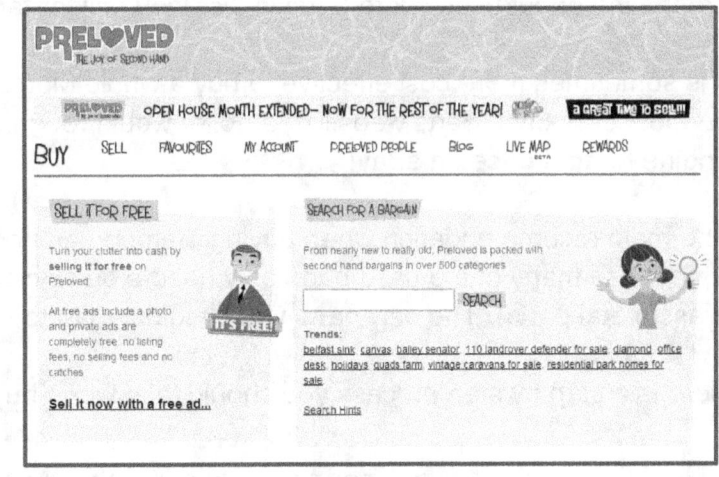

Preloved is one of the best websites out there to buy engines to restore (not for parts). Since using them, I have managed to get some great deals on engines, such as an SE3 for £30. The trick with Preloved is to keep updated with them, around once a week. If there is an engine that looks good to buy, you will then have to contact the seller about wanting to buy the engine. Since the seller sets the price for the engine, if they do not have much experience in Mamod engines, they could potentially underestimate the value of the engine. Considering you are now knowledgeable on Mamod engines ranging from 1936 to present, you can use your insight into Mamod engines to haggle even further with the seller to agree a price that is a real bargain.

There is an element of trust in Preloved, as it does not use a feedback system. That said, if you pay with PayPal, you are protected with PayPal's buyer protection scheme.

Ebay

Pros

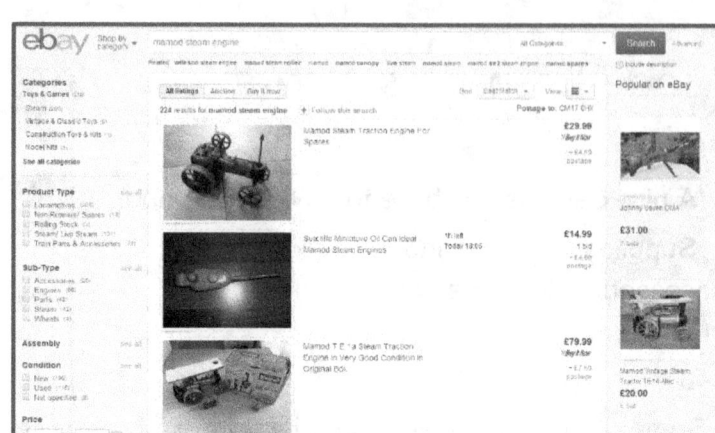

+ Many websites selling Mamod parts online also have eBay accounts linked to the websites (makes eBay a good place to compare the different website's pricing).

+ Huge range of parts from websites and independent sellers.

+ Usually quite cheap.

Cons

- Some sellers do not describe parts accurately.

- Auction prices for parts may be quite high due to many bidders and high competition.

Ebay is a great place to buy Mamod parts because most websites have an eBay account. Therefore, it is much easier to find the cheapest part if they are all listed on eBay. As well as this, there is the tragedy that some people buy Mamod engines to take apart and sell as spares. This does mean that there are lots of used spare parts for sale, which, if they are slightly dirty, can be cleaned to mint condition. The great thing about used spare parts is that they sell for far cheaper, even with postage included.

It is sometimes more cost effective to buy a whole Mamod engine than to buy new parts. For example, on most of Mamod parts websites, a boiler would roughly cost £20-£30 - you can buy a used Mamod engine for that price on eBay!

The main recommendation when buying Mamod parts online is to always try to buy used parts first. This is because many of the used parts only need a quick clean to make them look new again (such as brass/metal parts). The only parts you should try not to buy used are decal and fuel.

Here is a step by step process you should take when buying Mamod parts online:

1. Identify what parts you need to buy to restore your engine - list these parts to make purchasing and budgeting easier.

2. Surf eBay to try and buy these parts in used condition. If you cannot buy these parts used, or at a reasonable price, try the websites I have listed in this chapter.

3. Add to cart/basket all the parts you need to buy for each website and see which website works out the cheapest. Buy all the parts on that website since postage for all the parts will be combined making it cheaper to purchase the parts all on one website.

(Here are what I believe to be the four main websites for Mamod engines online although there will be many more you can find on the internet through searching on search engines. These include MamodOnline.co.uk, CJWSteam.com, ChaSteam.com, SteamCollector.com, SteamReplicas.co.uk, DreamSteam.co.uk and more which I advise you to check out too)

Glossary

A

Aesthetic: Concerned with beauty.

Accessories: Additional Mamod products which can be connected to Mamod engine for multiple uses:

- **Grinder**: A machine used to grind a surface.
- **Hammer**: A machine which lifts and releases a hammer.
- **Polishing mop**: A machine which has two polishing mops.
- **Power press**: A machine with a press that pushes down onto a flat surface.
- **WS1**: A workshop with a number of Mamod accessories mounted onto a base.

Annealing: The process of heating metal (or glass) and allowing it to cool slowly to remove internal stresses and, at the same time, toughen the metal.

Axle: A rod/spindle which is either fixed, or rotating, that passes through the centre of either a wheel or a group of wheels.

B

Base: A large rectangular sheet of metal on which all of the stationary engines are mounted.

Boiler: The brass container which holds water to be heated to generate pressurised steam.

Boiler band: A steel band coated in chrome which is used to secure the boiler on top of the firebox for the MM and SE range.

Brass: A yellow alloy of copper and zinc which tarnishes when in contact with oxygen (does not rust) and is self-lubricating.

Burner: A tray/container used to put fuel in so that the fuel can burn safely:

- **Fuel tray burner**: Used for burning solid fuel tablets.
- **Gas burner**. Used for burning gas from a pressurised tank of either butane or propane.
- **Vaporising burner**: Used for burning liquids, such as methylated spirit. The vaporising burner consists of a container filled with gauze covered by a mesh. The liquid is soaked into the gauze and burns above the mesh of the container.
- **Wick burner**: Used for burning liquids such as methylated spirit. The wick soaks up the fuel and steadily burns.

Burrs: A sharp edge produced as a result of cutting or machining a metal.

C

Canopy: An accessory to the Steam Tractor, Steam Roller and Steam Wagon which provides cover for the engine.

Canopy support: A long piece of metal connected at the back of the canopy and rear of the engine used to support the canopy at the rear of the engine.

Capillary action: The property of a liquid to seep into tiny gaps e.g. solder joints.

Catalyst: A substance which increases the rate of a chemical reaction without undergoing any chemical change.

Chassis: The internal framework supporting the engine (much like the skeleton of an animal).

Chimney: A vertical tube which directs wasted used steam from the cylinder into the atmosphere.

Chrome: Chromium is an element that is a hard white metal used in alloys as a decorative piece and provides a protective finish to parts.

Chromebox: A chrome plated steel enclosure, on mobile engines, to protect the fuel when burning.

Chrome housing/panel: A chrome plated steel rigid casing, found on the SP range that encloses and protects the boiler.

Closed system: A volume which does not gain, or lose matter, with its surroundings.

Connecting rod (on the lumber wagon): A long rod that connects two moving parts being the front and rear ends of the wagon trailer.

Conqueror: Mamod's second production boat which was powered by a FROG Revmaster electric engine.

Copper: A single element metal that is red/brown in colour.

Corrosion: The destruction and damage caused by oxidation (otherwise known as rust).

Crankshaft: A shaft that is connected to the end of the piston rod so that the oscillations of the piston can be converted into a rotating motion.

Cylinder: The tube-like piece of brass in which the piston moves back and forth.

D

Decal: A design that has been prepared on special paper for durable transfer onto another surface.

Dislocation: When the crystalline structure is disturbed causing a permanent partial shift.

Double acting slide valve cylinder: An assembly which has a sliding valve enclosed next to a cylinder which regulates and controls the timing of the steam that enters the cylinder. This control of steam causes the piston to oscillate.

Drive band: A thin, spring-like metal loop used to connect the flywheel to either a wheel on the engine, which is in contact with the floor (so the engine can move), or a Mamod accessory.

Ductile: The ability to be drawn out into a thin wire.

Dynamo: A machine that converts mechanical energy into electrical energy through rotating coils of copper wire in a magnetic field.

E

Emery paper: Stiff paper coated with powdered abrasive. The roughness of the emery paper is graded by grit per square inch (e.g. 200 is very course whilst 1,000 is a much finer grade).

Enamel paint: A paint that produces a hard wearing glossy finish.

End cap: The end panel of a Mamod boiler located at the front and back of the boiler. This is pressed onto the end of the boiler and soldered in place.

Engine bracket: A mount which holds and supports the moving parts of the engine.

Erosion: The action of being gradually worn away.

Exhaust gases: Gases that are emitted from an engine. For steam engines, steam is emitted from the cylinder to the chimney, whilst exhaust gases from the burning fuel dissipate into the atmosphere naturally.

F

Firebox: The chamber of the steam engine in which the fuel is burnt. It protects the flame from outside surroundings, such as wind and rain and enables the heat allowing the fire to reach higher temperatures.

Flux: A chemical cleaning agent used to clean the surface of a metal when soldering.

Flywheel: A rotating wheel with a relatively high mass to store rotational energy which enables the piston cycle to continue.

Forward/Reverse Lever: A lever on engines which changes the height and angle of the cylinder; adjusting which direction the cylinder rotates at (hence controlling the direction of the engine).

Fred Dibnah (MBE): An iconic English steeplejack and television personality who was well known for his keen interest in mechanical engineering, the Industrial Revolution and steam engines.

Friction: The resistive force occurring when two objects move over one another.

Front axle: A rod which passes through the centre of the two front wheels on mobile engines.

Front fork: The component which encloses the front axle and is attached to the leaf spring to connect the front of the engine to the front axle and leaf spring.

Fuel: Material that is burned to produce heat or power:

- **Gas fuel**: Gas fuel is burned by using the gas burner by Mamod. The usual gas used in gas burners is propane or butane.
- **Gel fuel**: A form of alcoholic gel that leaves residue when burnt.
- **Methylated spirit fuel**: Alcohol with an added 10% methanol, giving an invisible flame when burnt.
- **Solid fuel tablets**: Tablets made from pastilles carburant solide that are used for Mamod fuel.

G

Gearing: An arrangement of toothed wheels that engage with others to alter the speed of the piston and the speed of the flywheel.

Geoffrey Malins: Founder of MalinsModels, latterly becoming Mamod.

H

Hard soldered: A type of solder that contains silver and melts at a relatively high temperature, much higher than normal solder.

Heat resistant: The ability to resist excessive heat without failing.

Hobbies: The Company Geoffrey Malins worked for before he started MalinsModels.

Hub cap: A chrome coated piece of steel which is pressed onto the ends of the axles. The cap is kept in place with the use of tangs/barbs inside the part, to prevent the wheel from falling off the axle. Mamod also created rubber hubcaps too.

I

Ideal gas: A hypothetical gas which obeys the gas laws exactly (takes up negligible space and has zero interactions).

J

K

Kinetic energy: A type of energy a body can have through the means of motion.

L

Leaf spring: A type of spring which is made of a number of strips of metal curved slightly upwards and clamped together one above the other.

M

MalinsModels: The name of the British toy manufacturer in 1937 before it became popularly known as Mamod.

Mamod: British toy manufacturer specialising in manufacturing live steam models.

Mazak: Part of the Zamak family of alloys and made up of an alloy of magnesium, zinc and aluminium.

ME engine: A range of engines that enclosed the steam engine within a chrome housing with a propeller attached to the end of the crankshaft. There were four variations: the ME1, ME2, ME3 and the Meccano-styled MEC1.

Meteor: Mamod's first production boat which was considered a failure due to poor sales.

MM engine: A range of engines by Mamod that are, to date, the smallest engines Mamod have ever produced.

Mobile engine: An engine that has the ability to move, such as a road vehicle.

N

O

Open system: The opposite to a closed system, an area which can gain or lose matter with its surroundings.

O-ring: A type of seal in the form of a ring with a circular cross section that is typically made from rubber.

Oscillate: The motion of moving back and forth in a regular rhythm (*The piston oscillates*).

Oxidisation: The addition of oxygen to a compound. In simple terms, this is when a substance comes into contact with oxygen and causes the oxygen to become part of the molecular structure of the substance and is otherwise known as rust.

P

Piping: Long cylindrical pieces of copper to connect the boiler to the cylinder allowing steam to flow from the boiler to the cylinder.

Piston: A cylindrical body held inside the cylinder which moves back and forth due to the pressurised steam from the boiler.

Piston rod: A rod which connects the piston to the crankshaft.

Pressure: Continuous physical force exerted on, or against, an object by something being in contact with it.

Pulley: A wheel-like part which fits onto the crankshaft and has a groove so a drive band can connect the engine to a Mamod accessory.

Q

R

Regulator: Controls the rate of flow of steam from the boiler into the cylinder; hence, controlling the speed of the engine.

Rivet: A short metal pin used for holding together two plates of metal.

Rollers: Metal cylinders located at the front of the Mamod steam roller which are connected to and rotate on the same axis as the front axle.

RS engine: A range of Mamod steam trains that were able to run on 'O' gauge track and featured twin cylinders.

Rust: A reddish/yellowish brown flaking coating on either iron or steel that is caused by oxidation. Water works as a catalyst to speeding up the chemical reaction of rust.

Rust Converter: A chemical which treats the oxidisation of iron, or steel, which has formed on the metal by converting it into a protective chemical barrier.

S

SA engine: A range of Mamod models resembling roadsters, tourers and limousines from the past.

Safety valve: A valve that is located on top of all Mamod boilers which automatically opens to relieve excessive pressure.

Satin paint: A type of paint with a finish between glossy and matt.

Scuttle: A metal container used on a Mamod to attach to the back of a mobile engine and holds the burner. Traditionally, a scuttle would be used to fetch and store coal on a steam engine.

SE engine: A range of stationary engines manufactured by Mamod from 1936-1979.

Sight glass: An assembly on the rear of some Mamod boilers which has glass pressed against the boiler with either screws or copper rivets and is used to indicate the water level inside the boiler.

Solder: A low melting alloy, especially one based on lead and tin, or silver which is used for joining less fusible metals.

Soot: A black powdery or flaky substance consisting mostly of carbon, which formulates by the incomplete burning of an organic matter (such as Mamod fuel tablets).

SP engine: A range of stationary engines manufactured by Mamod from 1979 to the present day.

SR engine: The Mamod steam roller is a mobile engine with rollers at the front to replicate a steam roller which would have been used to flatten road surfaces.

Stationary engine: An engine that remains in a fixed position without the ability to move.

Steel: An alloy of iron and carbon which is a dull grey in colour.

Steering rod: A long rod which connects to either the front of the steam roller or down the end of the traction engine's chimney to control the direction the wheels face.

'Supercharge' or 'Superheat': The ability to increase the output power of a Mamod engine by looping the copper piping (to the cylinder) under the boiler so it is heated by the fuel, making the steam in the piping hotter and at an increased pressure (can exert a larger force).

SW engine: An engine designed by Steve Malins. The SW1 steam wagon resembles steam wagons which were road vehicles that were used for carrying freight.

T

Tarnish: The dullness of colour and loss of brightness, especially due to the exposure to air and/or moisture. Brass oxidisation is known as tarnishing since the colour of brass fades and becomes darker and duller.

TE engine: A mobile engine by Mamod which replicates steam traction engines used to pull heavy goods.

Thread: A manufactured spiral in male or female form which enables parts to engage which each other (the safety valve, water level plug and whistle are screwed into threads that are in the boiler).

Torque: A force that causes rotation.

U

V

W

Wagon: A vehicle designed to be pulled by an engine and is used for transporting goods such as wood and fuel.

Washer: A small flat rubber ring fixed between two joining surfaces to spread the pressure and act as a seal.

Water level plug: A screw-in plug located at the rear of the older Mamod boilers which is used to prevent overfilling of the boiler with water.

Whistle: A part on a steam engine which, when forcing air through the small hole in the part, emits a clear, high-pitched sound.

White spirit: A colourless liquid that has been distilled from petroleum and is used as a paint thinner or solvent.
Warning: white spirit is irritant to skin and dangerous to the environment and must be used with due consideration and care.

X

Y

Z

www.ingramcontent.com/pod-product-compliance
Lightning Source LLC
Chambersburg PA
CBHW081048170526
45158CB00006B/1896